T0211995

Immanence and the Animal

This book reexamines the concept of the animal on the plane of immanence, as opposed to the traditional viewpoint founded on the plane of transcendence.

Following Deleuze and Guattari's notion that philosophy is a discipline of creating concepts, this book traces how the concept of the animal was created in the history of philosophy through re-reading the works of Descartes, Kant, Heidegger, Derrida and Levinas. Their theories show that the concept of the animal was constructed on the "plane of transcendence" as subservient to the self-serving human, who represents the animal as a negative entity devoid of reason, ethics, the ability to enter into political alliances or even die. With this perspective and a range of theories from thinkers such as Spinoza, Nancy, Haraway and Braidotti as the groundwork, a new positive concept of the animal, operating on the plane of immanence, is sketched out, compelling a reappraisal of the relationships between body and thought, ethics and politics, or life and death.

With comprehensive interpretations of the views of several key philosophers, from Kant and Heidegger to Deleuze, Derrida and Agamben, this book will be valuable for scholars of theoretical animal studies and continental philosophy interested in the philosophical significance of the animal question.

Krzysztof Skonieczny is Assistant Professor at the Faculty of "Artes Liberales," University of Warsaw. His interests include animal studies, political philosophy and the question of atheism. He coedited (with Szymon Wróbel) *Atheism Revisited. Rethinking Modernity and Inventing New Modes of Life* (2020).

Routledge Human–Animal Studies Series
Series edited by Henry Buller
Professor of Geography, University of Exeter, UK

The new *Routledge Human–Animal Studies Series* offers a much-needed forum for original, innovative and cutting-edge research and analysis to explore human–animal relations across the social sciences and humanities. Titles within the series are empirically and/or theoretically informed and explore a range of dynamic, captivating and highly relevant topics, drawing across the humanities and social sciences in an avowedly interdisciplinary perspective. This series will encourage new theoretical perspectives and highlight ground-breaking research that reflects the dynamism and vibrancy of current animal studies. The series is aimed at upper-level undergraduates, researchers and research students as well as academics and policy-makers across a wide range of social science and humanities disciplines.

Carceral Space, Prisoners and Animals
Karen M. Morin

Historical Animal Geographies
Edited by Sharon Wilcox and Stephanie Rutherford

Animals, Anthropomorphism, and Mediated Encounters
Claire Parkinson

Horse Breeds and Human Society
Purity, Identity and the Making of the Modern Horse
Edited by Kristen Guest and Monica Mattfeld

Immanence and the Animal
A Conceptual Inquiry
Krzysztof Skonieczny

For more information about this series, please visit: www.routledge.com/
Routledge-Human-Animal-Studies-Series/book-series/RASS

Immanence and the Animal
A Conceptual Inquiry

Krzysztof Skonieczny

Routledge
Taylor & Francis Group

LONDON AND NEW YORK

First published 2020
by Routledge
2 Park Square, Milton Park, Abingdon, Oxon OX14 4RN

and by Routledge
605 Third Avenue, New York, NY 10017

Routledge is an imprint of the Taylor & Francis Group, an informa business

First issued in paperback 2021

Publisher's Note
The publisher has gone to great lengths to ensure the quality of this
reprint but points out that some imperfections in the original copies may
be apparent.

British Library Cataloguing-in-Publication Data
A catalogue record for this book is available from the British Library

Library of Congress Cataloging-in-Publication Data
A catalog record for this book has been requested

ISBN 13: 978-0-367-43720-6 (hbk)
ISBN 13: 978-1-03-223821-0 (pbk)
ISBN 13: 978-1-003-00528-5 (ebk)

Typeset in Times New Roman
by Apex CoVantage, LLC

Contents

Acknowledgments

This work is based on research that was supported by the Foundation for Polish Science International PhD Program, co-financed by the European Union within the European Regional Development Fund. This book started as a PhD project completed during the International PhD Program "The Traditions of Mediterranean Humanism and the Challenges of Our Times: The Frontiers of Humanity" at the Faculty of "Artes Liberales," University of Warsaw. It has certainly been influenced – or indeed shaped – by the vibrant and stimulating interdisciplinary milieu of the Faculty, where I have been privileged to participate in a number of fascinating seminars both as a PhD candidate and, later, as assistant professor. I owe gratitude to everyone with whom I crossed paths at this exceptional institution.

In particular, and first of all, I would like to thank my advisor, Professor Szymon Wróbel, for his enlightening and challenging mentorship during the preparation of my dissertation, and for all the help and collaboration afterwards. I am immensely grateful to Professor Jerzy Axer, the Dean of the Faculty and, later, the Director of the Collegium Artes Liberales, for all the opportunities I have had as a PhD candidate and an employee. Professor Jan Miernowski's superhuman work ethic has been a constant source of inspiration. Special thanks are due to the reviewers of my PhD thesis, Professor Monika Bakke and Professor Tadeusz Sławek; their helpful comments and tough questions helped pave the way for this book.

I owe a great deal to my colleagues from the PhD program and "Artes Liberales," especially Olimpia Dragouni, Katarzyna Szafranowska, Paweł Miech, Ewa Janion, Natalia Wawrzyniak, Kamil Wielecki and Julia Lewandowska, for their helpful comments and help in many circumstances, from the loftiest to the most mundane.

Much of the initial work on this project has been completed during my stays at the Department of Comparative Literature at SUNY Buffalo, led by professor David E. Johnson, and the Centre d'Études Supérieures de la Renaissance in Tours, led by professor Philippe Vendrix. I would like to express my gratitude to those institutions for hosting me as a Visiting Scholar. I especially thank Professors Ewa Płonowska-Ziarek and Krzysztof Ziarek for their help and generous hospitality in Buffalo.

I am grateful to the Routledge Human–Animal Studies series editor, Henry Buller, for giving this book a chance, and to Faye Leerink for patiently guiding me through the intricacies of the submission process.

Special thanks go to Julia Cydejko for her relentless support during the completion of this project.

Introduction

This book is not about animals.

The concept of the animal, which is the main focus of this book, does not – as I will try to show in the following chapters – strictly coincide with what we could call actual (or "empirical") animals. It is a specifically philosophical creation and should be treated as such, and only through a thorough analysis of how this concept works – how it is created, how it changes, what its historical incarnations are and how we can re-create it – can we begin to understand how it influences the way how we think about and treat animals.

Many parts of this book may therefore seem a very abstract undertaking. This might disqualify it in the eyes of those with a more practically oriented view of animal studies as a field, which sees its first and most important goal to be the immediate amelioration of the fate of the billions of animals suffering and dying every year as a direct or indirect result of human action. These endeavors were indeed the cornerstone of the field, presented in books such as Peter Singer's *Animal Liberation*, Tom Regan's *The Case for Animal Rights* or texts written by Gary Francione, Martha Nussbaum or Paola Cavalieri.

Equally important are interventions that seek to see animals in a wider, political frame, arguing that individualistic, purely ethical frameworks put forth by some of the authors mentioned previously amount to little more than wishful thinking, as the only way to amend the way people treat animals is to change the way they are treated in the current socio-political context, which, in turn, cannot be accomplished without a radical change of this context, i.e., without serious changes in the capitalist mode of production or indeed abolishing capitalism altogether. Strong voices in this approach – sometimes dubbed *critical* animal studies, as it merges caring about animals and cultural criticism – include Simon Wise, Nicole Shukin and, to a certain extent, Cary Wolfe.

Both of these strategies have proven ethically and politically successful, resulting in concrete initiatives such as *The Great Ape Project* and influencing large numbers of people around the world to change their dietary habits and whole lifestyles. I feel that as this book is often critical of the theories and arguments put forth by those thinkers, it is important to note that their contribution to the field – and to animal welfare in general – is invaluable.

Remaining indebted to these thinkers, this study has a different, perhaps more fundamental goal, whose effects however will most probably be much more modest. Aiming to rethink the concept of the animal – as I hope to show – leads to an attempt to rethink notions such as thinking, ethics, politics and death through a collapse of a dominating conceptual framework, exemplified in this work by familiar figures such as Aquinas, Descartes or Kant but also their philosophical opponents such as Freud, Levinas or Agamben. Indeed, I am hoping to show that the rethinking of the concept of the animal I propose is an act akin to removing a capstone from an arch – lead as it may to a conceptual catastrophe, it will hopefully force us to engage in meticulous reconstruction.

Such work is, of course, not without allies. Among the many theorists referred to in this book, perhaps the most prominent place is held by Gilles Deleuze. However, contrary to most treatments of Deleuze and the animal question, I almost entirely ignore the notion of becoming-animal and other explicitly animal themes in the French philosopher's work. The reason for this is not only the extensive treatment of becoming-animal in the already-extant literature, but especially the fact that I believe that by singling out this notion, we fail to take full advantage of the possibilities given by Deleuze's philosophy with regards to the concept of the animal. This point merits longer clarification.

Matthew Calarco classifies Deleuze as one representing the "indistinction" approach with regards to the difference between man and animal. In Calarco's own words, this approach

> stakes itself on the [. . .] wager [. . .] that it is possible to invent new modes of thought beyond the human/animal distinction. [. . .] [It] proceeds from a space in which supposedly insuperable distinctions between human beings and animals into a radical *indistinction* and where the human/animal distinction (in both its classical and more complicated deconstructive form) no longer serves as a guardrail for thought and practice.[1]

This approach thus entails finding a common space for considering men and animals which would render the traditional division irrelevant. In case of Deleuze, an example of such a common space is what he calls a "zone of indiscernibility."

In the book on Francis Bacon, from which Calarco draws the term, Deleuze gives two main examples of zones of indiscernibility. The first one is the body:

> [The] objective zone of indiscernibility is the entire body, but the body insofar as it is flesh or meat. Of course, the body has bones as well, but bones are only its spatial structure.[2]

And then, about a page later: "Meat is the common zone of man and the beast, their zone of indiscernibility."[3] Insofar as man and animal alike are meaty creatures, creatures of the flesh, a zone of undecidability can be found between them; they can enter a process of becoming that would make them indiscernible (one could ask, of course, if a coral or a jellyfish can be classified as "meaty" enough).

The bones are rejected from the zone, for they form a structure – this thread is very similar to what Deleuze, in other texts, says of the organism. Becoming-animal consists in shedding the three "great strata";[4] if one becomes-animal (i.e., enters a zone of indiscernibility) one stops not only being human, but also an organism.

The second zone of indiscernibility is the head:

> As a portraitist, Bacon is a painter of heads, not faces, and there is a great difference between the two. For the face is a structured, spatial organisation that conceals the head, whereas the head is dependent on the body, even if it is the point of the body, its culmination. It is not that the head lacks spirit; but it is a spirit in bodily form, a corporeal and vital breath, an animal spirit. It is the animal spirit of man: a pig-spirit, a buffalo-spirit, a dog-spirit, a bat-spirit. . . . Bacon thus pursues a very peculiar project as a portrait painter: *to dismantle the face*, to rediscover the head or make it emerge from beneath the face.[5]

The head/face distinction is dear to Deleuze, and his critical approach to the face is clearly directed at Levinas, for whom the face is distinctive of the human. According to Deleuze and Guattari the face is not a primordial, naked phenomenon but, on the contrary, a construction imposed on us by the socioeconomic (metaphysical?) system of the West. It is a mix of the other two great strata that make us "human" – subjectification and signification. It is therefore only by "losing face" that one may embark on the process of becoming animal, that one can enter the zone of indiscernibility.

In the end, the two elements (the head and the flesh) are essentially akin, strange as it may seem:

> But can one say the same thing, exactly the same thing, about meat and the head, namely, that they are the zone of objective indecision between man and animal? Can one say objectively that the head is meat (just as meat is spirit)? Of all the parts of the body, is not the head the part that is closest to the bone? [. . .] Yet Bacon does not seem to think of the head in this manner. The bone belongs to the face, not to the head. [. . .] The head is deboned rather than bony [. . .] The head is of the flesh.[6]

The head, in opposition to the face, is closer to meat than to the organism. Faceless, far from forming the main locus of human identity, it becomes a non-privileged part of the flesh-body.

At first glance (and that seems to be Calarco's opinion), this approach succeeds in breaking the man/animal division and making the human and the animal meet on a "neutral territory." It thus manages to evade the mistakes of "humanizing" the animal or "animalizing" the human (I further comment upon these mistakes in the chapter "False Immanence"). However, from the perspective of this text, which finds it essential to analyze the very border between the animal and the human, it seems worthless for that very reason: an introductory reading suggests it does away with the difference, rather than offer an answer to the preliminary question concerning where it came from.

In fact, the opposite is the case. The previous appraisal of the indistinction approach is based upon its "final product," where the boundary has already been destroyed or crossed. However, the process leading up to it can prove to be much more problematic.

Although Deleuze does not show us this method in a step-by-step fashion, but rather follows a logic which flows from Bacon's paintings, the logic of his analysis is quite clear. Both of the zones of indistinction (faceless heads and boneless flesh) are the result of a particular movement which is enabled by the existence of a particular frontier between the animal and the human. A very classical frontier, it needs to be added.

In the first case Deleuze is dealing with the body/mind (or body/spirit) distinction. Classically, the body is on the side of the animal, while the spirit is on the side of the human. This is a paradigmatic Cartesian moment.[7] Entering a zone of indistinction, which is composed of both becoming-animal of the human and becoming-human of the animal, means not so much crossing the barrier as treading the line of undecidability that forms it. In this way, the spirited, bodiless human becomes meaty, animal flesh, but at the same time the animal flesh becomes a spirited body.

Roughly the same process is at work in the second aforementioned zone, where the human becomes faceless – even though Deleuze does not tell us the "animal" side of the story. As indicated before, Deleuze's focus on the face is clearly directed against Levinas. For the author of *Totality and Infinity*, the experience of the naked face of the Other creates in us a form of unconditional ethical response, which, in its minimal form, creates a prohibition of murder. This way, any time we recognize someone as having (or being) a face, we are forced to think of him as a subject of ethics, one we have responsibility for. This ethical community has the same extent as the human community. Only humans have faces.[8] Animals just have heads. Thus, by defacing himself, man stops being human and enters into a realm of the animal.

The main problem is that for this strategy of breaking the barrier to work, one needs to believe in the existence of the barrier itself. Without acknowledging that man is in fact a being which has a face, and the animal is one which has not got one, one cannot possibly say that becoming-animal involves a losing-face. Without acknowledging that the human is a spirited being while the animal is defined by just having a body, it is impossible to locate the zone of indiscernibility between the human and the animal in spirited, boneless flesh.

This is why the notion of becoming-animal is of little use of this work. It is certainly useful or inspiring in a practical or political sense, providing guidelines for subversive or revolutionary action – it is for these reasons that Calarco values it so highly, and it is in this respect that, for example, Donna Haraway finds it lacking. However, it starting as it does from a very classical understanding of the caesura between man and animal, becoming-animal cannot in itself be used to put those concepts into question and see the basic logic operating behind the difference between them.

Even so – as I indicated earlier – I believe that a way to understand the fundaments of the human–animal divide can be found in Deleuze's understanding of what philosophy is, namely the practice of creating concepts on planes of transcendence or immanence; I thoroughly analyze this approach in Chapter 2, where I also show how it changes the shape of philosophical polemics as well as the relationship between philosophy and its Others, especially science.

However, there is one merit to Deleuze's becoming-animal approach that I would like to comment on. As said earlier, Deleuze starts his revolutionary theorizing from a very classical understanding of the animal and the human, only to find ways to subvert it. This strategy is in direct contrast with the many "posthumanist" theorists who claim that, for example, through the "new proximity to animals, the planetary dimension and the high level of technological mediation"[9] our subjectivity has changed so much that we can no longer be properly called "human" – or indeed that what is known as "human" never really described the actual reality of what "human beings" are. While this work remains deeply theoretically indebted to such authors as Rosi Braidotti or Donna Haraway – as well as their inspirations, among whom Spinoza plays a crucial role – I believe that Deleuze's approach would see the posthuman condition not as a fact, but rather as a task. The theoretical challenge, then, is not to see that "we have never been human," but rather to understand that despite the changes that should wring us out of our modernist (or Christian, or Greek) humanity, we have remained "human, all too human."

Notes

1 Matthew Calarco, "Identity: Difference. Indistinction," *CR: The New Centennial Review* 11, no. 2 (2011): 54. See also Matthew Calarco, *Thinking Through Animals: Identity, Difference, Indistinction* (Stanford: Stanford Briefs, 2015).
2 Gilles Deleuze, *Francis Bacon: The Logic of Sensation*, trans. Daniel W. Smith (London and New York: Continuum, 2004), 22.
3 Deleuze, *Bacon*, 23.
4 See Gilles Deleuze and Felix Guattari, *A Thousand Plateaus*, trans. Brian Massumi (Minneapolis: University of Minnesota Press, 2005), esp. p. 159: "Let us consider the three great strata concerning us, in other words, the ones that most directly bind us: the organism, significance, and subjectification. The surface of the organism, the angle of significance and interpretation, and the point of subjectification or subjection. You will be organized, you will be an organism, you will articulate your body – otherwise you're just depraved. You will be signifier and signified, interpreter and interpreted – otherwise you're just a deviant. You will be a subject, nailed down as one, a subject of the enunciation recoiled into a subject of the statement – otherwise you're just a tramp."
5 Deleuze, *Bacon*, 20–21.
6 Deleuze, *Bacon*, 25.
7 I analyze how the body–animal/spirit–human distinction works in Descartes' writings in Chapter 3.
8 A more thorough analysis of Levinas' approach to the man/animal difference can be found in the Chapter 4.
9 Rosi Braidotti, *The Posthuman* (Cambridge: Polity Press, 2013), 94.

1 Philosophical zoology
Why philosophers still believe in the human–animal caesura

As Matthew Calarco writes in the context of Jacques Derrida's work, "[t]he human/animal distinction is one of the most saturated and exhausted distinctions in the Western metaphysical tradition,"[1] and therefore the only way to rethink the concept of the animal would require us to "invent other concepts, groupings and frameworks."[2] And yet, as I will try to show on the pages that follow, any such reinvention will have to proceed through a rethinking of the distinction in question.

Why is the animal, especially for a philosopher, such a special, strange and perplexing "word"[3]? Derrida explains:

> Confined within this catch-all concept, within this vast encampment of the animal, in this general singular, within the strict enclosure of this definite article ("the Animal" and not "animals"), as in a virgin forest, a zoo, a hunting or fishing ground, a paddock or an abattoir, a space of domestication, are all the living things that man does not recognize as his fellows, his neighbors or his brothers. And that is so in spite of the infinite space that separates the lizard from the dog, the protozoon from the dolphin, the shark from the lamb, the parrot from the chimpanzee, the camel from the eagle, the squirrel from the tiger, the elephant from the cat, the ant from the silkworm, or the hedgehog from the echidna.[4]

One must not misread Derrida's statement as simply indicating that the term "Animal" is too wide to still serve as a valid notion, be it in philosophy or in common language. Indeed, there are many terms that are much wider – think of "being," which would include all animals as well as inanimate objects: simply everything that *is* – and are still considered valid and useful. No, the wide array of beings that are included within this "catch-all concept" is only perplexing when we realize that there is one being that is excluded from it (and it is precisely the one that does the excluding) – man.[5]

This statement might seem strange on quite a few levels. Etymologically – and following the Aristotelian tradition – any being capable of moving on its own should be called an animal. Biologically, man is as much an animal as an ape or a cat and is much more closely related to an ape than a cat is. Even classical philosophical definitions of man seemingly seek to include him in the animal realm. For example, in *animal rationale*, man (species) is an animal (genus) that is rational (specific difference), in contrast to others, that are not rational, but remain in the same genus.

However, if this indeed were the case, one of the most persistent philosophical questions – what is the difference between animals and men, between the human and the animal? – would reveal itself to be illogical. One does not ask about the difference between a genus and one of its species (one might ask, what differentiates one species from others in the same genus, and this is precisely the question which a classical definition seems to address). We would not treat the question "What is the difference between a horse and an animal?" as a valid philosophical inquiry, regardless of our interest in horses.

And yet, when it comes to man, asking the question is not only regarded as philosophically sound, it is also rooted in a deep conviction that the difference between man and animal is more radical or "abyssal" (Derrida again) than it would seem fit for two species of the same genus. Its persistence in philosophical texts is as puzzling as the fact that it is rarely formulated explicitly, as if it was too obvious to mention. This is perhaps most strikingly visible in Descartes' *Letter to the Marquess of Newcastle*, where the French philosopher explains his views about the possibility that animals can think. Having granted that the similarity between the organs of animals and humans is striking, he admits one could hypothesize that since those organs allow man to think, animals should also be able to think, even in a less perfect manner. However, in Descartes' metaphysics, in order to properly think, a being would need to possess an (immortal) soul. And so, as the author of the *Discourse on the Method* concludes:

> This is unlikely, because there is no reason to believe it of some animals without believing it of all, and many of them such as oysters and sponges are too imperfect for this to be credible.[6]

This is clearly an example of *reductio ad absurdum* – and the only instance in which it works as a rhetorical tool is when the author of the letter is convinced that both him and his addressee share the belief that all animals have a common essence which is radically different from that of man. An ape is as much an animal as is an oyster; man, however, is of a totally different kind. In this view, an ape has more in common with an oyster than with a man.

This belief is far from an idiosyncrasy of Descartes'. Heidegger, for example, says that his thesis about animals being poor in world, which I will analyze more closely in the chapter "Anthropology,"

> does not tell us something merely about insects or merely about mammals, since it also includes, for example, non-articulated creatures, unicellular animals like amoebae, infusoria, sea urchins and the like – all animals, every animal.[7]

A similar tendency can be found in the often Aristotle-inspired philosophies claiming that man is the only one possessing a rational (and not only a vegetative and animal) soul, notably in the work of Aquinas; in Nietzsche's famous statement that mankind "is a rope fastened between animal and overman";[8] in Agamben's theses about how man differs from (all) animals because he undergoes a period of infancy; in Lacan's remarks about an animal's incapability of covering its tracks; or in Deleuze and Guattari's concept of becoming-animal. This list is – of course – far from complete, and does not indicate a wish to claim that there are no crucial differences between the philosophers it mentions regarding the theme of the animal, but it suffices to show that the belief in an existence of such an essential, radical frontier between man and animal – even when it meant that a given philosopher could say, almost on one breath, that man is (for example) a rational animal, and yet not an animal at all – has been shared by philosophers – at least in the Western world – regardless of nationality, historical period or creed.

Where does this inconsistent and illogical, and yet so pervasive, manner of thinking come from? A few answers come to mind. Firstly, philosophizing in terms of a radical human–animal caesura is a remnant of pre-Darwinian thinking. According to this view, one cannot really blame pre-Darwinian thinkers for this, for they simply did not know better; and conversely, anyone thinking this way after the theory of evolution had been widely accepted is simply wrong and uneducated in biology.

Both of these assumptions are at the least problematic. The theory of evolution was present well before Darwin – in fact, thinkers such as Aquinas or Descartes already toyed with similar ideas, and by the 19th century it was commonly accepted by biologists. As books concerned with the history of biology never cease to remind us, Darwin did not invent evolution, but found an explanation for how evolution happens, namely by means of natural selection. What is more, one does not need the Darwinian theory evolution to see the uncanny resemblances between man and other animals – in another fragment of the already quoted letter, Descartes admits that "the organs of their bodies are not very different from ours";[9] early-modern thinkers such as Montaigne and Charron famously said that "there is a greater distance between this man and that one than between this man and that beast";[10] and finally, biologists and taxonomists found it almost impossible to distinguish man from apes (the same apes that were, for the likes of Descartes, closer to sponges). As Linnaeus put it:

> I must remain in my workshop and consider man and his body as a naturalist, who hardly knows a single distinguishing mark which separates man from the apes, save for the fact that the latter have an empty space between their canines and their other teeth.[11]

What is more, *On the Origin of the Species* seems to have had a very limited impact on philosophy – strangely enough, more so in the 19th than the 20th century.[12] This problem cannot be explained away by invoking an alleged lack of biological education. We have seen that Nietzsche, an admirer of Darwin's, thought (at least in some points of his oeuvre) in terms of the radical human–animal cae-sura. So did Heidegger, whose remarks in *The Four Fundamental Concepts of Metaphysics* and elsewhere clearly show he was very much up to date with the achievements of the sciences of his time. The same tendency can be found in philosophical anthropologists such as Cassirer, who uses the age-old difference between (animal) reactions and (human) responses,[13] and introduces the term "*animal symbolicum*" to show the radical nature of the man–animal caesura – and all that in books informed by a wide and impressive reading of authors from vari-ous fields of the natural sciences and psychology.

In other words, the thought or evidence – or both – that man may be closer to animals than the radical nature of the purported difference between the two sug-gests, were there for the taking, but something stopped the philosophers in ques-tion in their tracks. Not that there is a need for philosophy to heed to scientific theories; in fact, Heidegger and many others would probably agree with Wittgen-stein – who thus also enters this list – that "[t]he Darwinian theory has no more to do with philosophy than any other hypothesis of natural science."[14] But there are scientific hypotheses, and there are – to paint with as broad a brushstroke as possible, which will be refined in the chapter "Polemology" – ways of thinking, and these tend to travel between the sciences: why was Descartes so enthusiastic to use William Harvey's discovery of the blood circulation system to prove his body–machine theory, but failed to take the striking similarities between human and animal brains as evidence of thought? Why did Heidegger so eagerly use von Uexküll's theories to show that animals have an environment/*umwelt* (which, as we will see in the chapter "Anthropology," Heidegger understood as a philosophi-cal statement, not just a biological one), but did not relate this to humans? Why does Agamben quote Linnaeus so favorably, and then proceeds to think in terms of the human–animal caesura, as we will see in the chapter "Anatomy"?

Again, a more profound determination seems to be at work here, something that is so deeply ingrained into (philosophical) thinking about man and animal that it allows or forces to uphold the previously cited logical inconsistencies, neglect observation and seemingly reject a scientific consensus. A possible sus-pect is religion – the Judeo-Christian tradition has long been blamed for the bad fate of animals, resulting from a clear-cut distinction between Adam and the rest of Creation, and the commandment to "have dominion over the fish of the sea, and over the fowl of the air, and over every living thing that moveth upon the earth."[15] We will later see that this is not entirely untrue, but for now, it remains problematic, at least as a straightforward and unqualified statement. From Saint Francis of Assisi to Andrew Linzey, there have been a great number of Christians who challenged this traditional stance on the basis of Christianity itself.[16] The famous gesture of Diogenes, who "plucked a fowl and brought it into the lecture-room with the words: 'Here is Plato's man,'"[17] upon hearing the definition of man as a "featherless biped," also seems to prove that the ancient Greek philosopher thought it absurd to define man in such proximity to animals[18] – and all that from a thinker who lived four centuries before Christ. In Hellenistic and later philosophy, it is the Stoics who are usually cited as those who most strongly opposed animal and man.[19] And even if we disregard the speciesist approaches outside Christianity, to even include Aquinas, Descartes, Nietzsche and Heidegger, not to mention Deleuze, Guattari and Lacan, in the same tradition (Judeo-Christian or any other), one would have to describe this tradition so broadly that it would have no explanatory value whatsoever.

The same seems to go for the notion of "metaphysics," which has been used as a supposed explanation for the manner of thinking in question, for example by Leonard Lawlor, who in his otherwise excellent book on Derrida's approach to the "animal question" speaks of "metaphysical separationism"[20] as one of the risks involved in speaking of the man–animal difference. Such nomenclature is at least misleading, because while the caesura between man and animal is certainly present in (at least some forms of) metaphysical thinking, saying that it itself is metaphysical, or – even worse – anyone thinking this way is *a fortiori* thinking metaphysically, gets us dangerously close to what could be dubbed "philosophical fear-mongering." Emotions aside, the most that can be said about the notion of metaphysical separationism, is that it seeks to *name* a phenomenon (i.e., thinking in terms of a radical caesura) rather than *explain* it, and thus is of little use at this point of the investigation I am pursuing.

Derrida's own notion of carnophallogocentrism, which he introduces in the 1990s to underscore that the metaphysical structure of power in the Western world is a sacrificial one and needs for its very existence the notion of non-criminal putting to death of animals and eating flesh,[21] also gives a name to the problem (perhaps an altogether different one than "metaphysical separationism") rather than seeks to explain what kind of thinking led to its appearing in the first place. To be fair to Derrida – and Lawlor as well – one must say that the stakes of their endeavors are different than mine: rather than find reasons for the existence of the caesura in the first place, they plan to tactically use it in order to put it in question and subvert it, eventually providing a "more sufficient response" to the question of the relationships between men and animals. In this, they are close to the Deleuzian becoming-animal approach I took pains to distance myself from in the Introduction. This does not mean that it is not a worthwhile undertaking, but the goal of the present book requires taking smaller steps.

What this also does not mean is that Derrida's approach is entirely useless to this work – on the contrary, his tireless escapades into animal–human borderlands can help make an important step to confirming, if not the root or logic of the difference itself, at least the terrain on which it is situated.

Derrida identifies the wider stakes of this strategy in the beginning of his discourse in *The Animal that Therefore I Am*. He is not interested in destroying the border between men and animals, bridging the gap, as it were, but rather – as I said before – wants to complicate it, show that the limit of what is "human" (and therefore what is "animal") is not as clearly demonstrable as we would perhaps like it to be. In his own words:

Everything I'll say will consist, certainly not in effacing the limit, but in multiplying its figures, in complicating, thickening, delinearizing, folding, and dividing the line precisely by making it increase and multiply. [. . .] So it will in no way mean questioning, even in the slightest, the limit [. . .] between Man with a capital M and animal with a capital A. [. . .] I have [. . .] never believed in [. . .] homogeneous continuity between what calls *itself* man and what *he* calls animal.[22]

A seemingly similar thread can be found in the interview with Elizabeth Roudinesco, *Violence Against Animals*:

> The gap between the "higher primates" and man is in any case abyssal, but this is also true for the gap between the "higher primates" and other animals.[23]

Both of these fragments say roughly the same thing: there is an abyssal gap between men and animals; however, there is an equally abyssal gap between different species of animals. This would seem to indicate that, according to Derrida, the difference between man and animal is not in any way special – since the difference between men and higher primates is as big as between the latter and other species. However, by capitalizing the two terms (Man and Animal), and by clinging to its importance, Derrida in fact keeps the human–animal frontier in place.

Based on such reconstructions of his thought, some critics claim to find an inconsistency in Derrida's thinking. One example is Matthew Calarco, whom I cited in the beginning of this chapter. For Calarco, the only reason to maintain the distinction between Man and Animal would be a strategic one – this, he notes, is a typical Derridian move, used for example in the speech/writing distinction or the case of the concept, when Derrida provisionally keeps certain notions if they are useful and/or perfectible. However, as evidenced by the quotation from the first paragraph of this chapter, Calarco finds no such use for this strategical maintaining of the animal/human distinction based on Derrida's writing.

Less sympathetic critics of Derrida, like Paola Cavalieri, make this point even more strongly, citing his condemnation of factory farming as radically inconsistent with the theses on the abyssal nature of the man–animal divide, and the fact that despite his promising stance against Heidegger's "humanistic bias,"[24] he stands firmly on the ground that animals are – at least from an ethical perspective, which is crucial for Cavalieri's approach – inferior to men. In the end, Derrida is classified as displaying a "mixture of concealed perfectionism, undeclared pro-human bias and confused rejection of naturalism."[25]

Rather than address these charges directly, I would like to propose a slightly different reading of Derrida's animal philosophy, one that will hopefully clear up the reasons behind his apparent inconsistencies, work towards showing his rejection of naturalism is far from confused, and assess the strategic use of the human–animal distinction, which will prove especially important for this investigation. For its value to be shown, it is Derrida's approach that in turn needs to be, "multiplied, complicated and divided." Doing so will include asking the question "who?" in the right way. This, in fact, is a Nietzschean way of questioning – and that of some of Socrates' interlocutors in Plato's dialogues – which replaces the classical "what?", the question of essence with a question of forces and stand-points or positions.[26]

Therefore rather than start by asking "what is an animal" or "what is the difference between man and animal," we should ask who Derrida is speaking *as*, or *from which position or standpoint* he is speaking, when he talks about The Animal/animals. This question needs to be answered in three different instances in order to reveal a logic that founds Derrida's apparent inconsistencies. First of all, we need to ask who is talking about animals, or whose discourse about animals we are talking about. Second, who Derrida (as the lecturer in *The Animal that Therefore I Am*) is in this exchange, what is the answer to the central question "who am I," or who he is following (double meaning of the French *qui je suis*, which Derrida often plays upon). And thirdly, we need to ask who is/are the animal/animals that is/are being talked about.

Derrida's most common strategy is the one dubbed the "difference" approach by Calarco. In this instance, the first question, "who is talking about animals?" question can be answered by identifying the "they" in the often-used expression "what *they* call animals." Clearly, Derrida himself is not one of "them." He says:

> I am saying "they," "what *they* call an animal," in order to mark clearly the fact that I have always secretly exempted myself from that world, and to indicate that my whole history, the whole genealogy of my questions, in truth everything I am, follow, think, write, trace, erase even, seems to me to be born from that exceptionalism and incited by that sentiment of election.[27]

"They" are therefore those who speak in a logic that allows them to pose the sharp and illogical caesura between the human and the animal.

And who is this animal that "they" so radically differentiate from man? It is mere fiction, an invented term, and what is more, a term that only serves man to talk about himself?

Who is Derrida here, then? In this structure he is the self-proclaimed elect of what "they" call animals, he speaks from the point of the *animot*, as the representative of the infinite differences between individuals, those who call themselves humans and those that they call animals alike.

The *animot*, a word sounding like the plural form of "animal" in French, but in writing containing the word "word" ("*mot*"), stresses both those differences and the fact that it is not a concept, but just a word to describe this strange combination of beings that are usually referred to as falling under the category of "the Animal." The *animot* is

> Neither a species nor a gender nor an individual, it is an irreducible living multiplicity of mortals, and rather than a double clone or a portmanteau word, a sort of monstrous hybrid.[28]

We therefore have the "them" who speak about the animal, the animal that is revealed as a fiction, and Derrida, who speaks from the position of the *animot*.

The second structure introduces an all-new animal, an all-new position of the speaker, and, so to speak, an all-new Derrida. It is best visible in the often – perhaps all too often – cited story of Derrida's cat, who enters the bathroom while his human companion is naked, stares at him for a moment, and immediately urges Derrida to open the door again to let him out. Derrida makes it clear that he is talking about a real, individual cat, not an allegory, not a myth or symbol, not Montaigne's cat and certainly not that of Descartes. As Derrida continues:

> [I]t is true that I identify it as a male or female cat. But even before that iden-
> tification, it comes to me as *this* irreplaceable living being that one day enters
> my space, into this place where it can encounter me, see me, even see me
> naked. Nothing can ever rob me of the certainty that what we have here is an
> existence that refuses to be conceptualized [*rebelle à tout concept*].[29]

This individual that enters Derrida's bathroom leaves the naked man in a state of complete passivity, unable to utter a single word. He leaves him ashamed, perhaps blushing, and unable to *conceptualize* the being he encountered. The long stream of negative delineations of what the cat is certainly not only strengthens this impression. This story can only be told afterwards, when Derrida's face returns to its natural color. And even then, the word "cat" or "animal" can be used only provisionally.

Both of these figures condition each other. It is the individuality of the "animal" (cat) that makes it impossible for the "man" (Derrida) to say anything about himself, to "properly" understand himself *as* man. A man, as defined within the strict animal–human caesura, should not feel ashamed when facing an animal.

This is the second structure. The animal is a truly individual being (I use the term "truly individual," because it is a rare occurrence, somewhat of a fortuitous encounter, but also because of how it cancels out the general term "animal"). Derrida is the individual – perhaps it should rather be said that the story happened to an individual who only later can identify himself as Derrida – put to a state of passivity. Derrida the speaker/philosopher appears only later, and is still unable to give justice to the encounter.[30]

The third structure is only touched upon, but it needs to mentioned to present Derrida's stance and to understand its strategic meaning to this text. It is easiest to spot in a few simple remarks in the beginning of *The Animal that Therefore I Am* which are uttered right before Derrida assumes his elect stance. While explaining his own approach of "complicating, thickening, delinearizing, folding, and dividing the line" (see previous) he also says:

> It will not be a matter of attacking frontally or antithetically the thesis of
> philosophical or common sense on which has been constructed the relation
> to the self, the presentation of self of human life, the autobiography of the
> human species, the whole history of the self that man recounts to himself,
> that is to say, the thesis of a limit as rupture or abyss between those who say
> "we, men," or "I, a human," and what this man among men who say "we,"

what he *calls* the animal or animals. I shan't for a single moment venture to contest that thesis, nor the rupture or abyss between this "I-we" and what we *call* animals.[31]

Even if he assumes this position only provisionally, Derrida is clearly speaking from a point of view that treats the human–animal caesura seriously as an important – if not *the most* important – factor constructing human identity. This "human identity" or "self" has been assumed by philosophers and accepted by those who use "common sense," who decided to call themselves "human" and to create the category of "animal" in order to name everything that is not them, but still is a living being. Derrida here is in the position of the *them* from the first structure – he believes that the human–animal caesura is crucial for the understanding of both terms of the division, even if it is rather an understanding of the former and the misunderstanding of the latter, or the creation of both (this is why he needs to assume the position of the *animot* in the first structure to cease to be "human" and see the absurdity of the division). He is also the one speaking about the animal. The animal itself is indeed again the one from the first structure – one that is different from the human in an abyssal way. Here, no positive explanation is necessary – everything that has been *classically* said about the animal (that it has no language, no reason, no politics, no ethics, etc.) is true about the animal from the third structure.

After the three structures of his discourse have been distinguished from one another and identified, we can see that Derrida seems less inconsistent than Calarco or Cavalieri suggest. When he claims not wanting to question the limit between "Man and Animal" (capitalized by Derrida), he simply admits that the only way one can use these two notions is by admitting that there is a radical gap between them, and only in this pair can they be understood. And of course, from a supposed outside viewer (the "elect" position of the *animot*), this division is imaginary and absurd. However, in such a situation we are left either talking about the "monstrous hybrid" that is the *animot*, or we are left speechless, standing before an individual which escapes any and all categorizations.

While Derrida himself does not address the question about the foundation of thinking in terms of a strict caesura between man and animal, carefully differentiating between these three possible ways of speaking – which are perhaps not the only ones, but serve to provide a useful scaffolding for what is to come – helps to determine the terrain on which one should move to find such foundations. The first structure shows that whenever viewed from outside, the classical, clear-cut frontier between man and animal is impossible to understand or even trace. In a certain way this echoes the doubts of biologists – or anyone who has ever seen an ape's or even a pig's behavior, not to mention their insides – who failed to see any essential difference between bodies of men and animals; it also confirms the logical intuitions according to which talking about the clear-cut caesura leads to inconsistencies. What is more, it explains those doubts and intuitions by showing that they are dependent on the outside position from which they try to evaluate the frontier.

Individual beings – as seen from the second structure – refuse to be easily pin-pointed to categories, and, as the example of the cat teaches us, an encounter with them leaves us unable to conceptualize their being; it leaves us in an utterly anti-philosophical state. When encountering a truly individual being, one is unable to understand oneself as man, and equally understand the being as animal.

The radical difference between man and animal is only understandable and logical from within, from the position of the *we* who distribute the names of man and animal, as described by the third structure. Therefore in order to under-stand its roots, one must first believe in its existence and subscribe to its logic, as illogical as it might seem, rather than try to evaluate it from an outside point of view – regardless of how tempting it might be to contemptuously watch its propo-nents from a biological or logical higher ground. Philosophical zoology operates according to its own laws, which, however not closed to scientific, religious or any outside discourse content-wise, always subdue any outside input to its form.

Any thought that seeks to unwork the difference needs to proceed by first under-standing those laws, or it runs the risk of being subjected to them. Many theoreti-cal approaches in animal studies seem to fall into this trap, especially those who subscribe to what Matthew Calarco calls the "identity" approach. These theorists

> seek to establish a [. . .] moral identity between human beings and animals most often through a rigorous application of Darwinian ontology on the one hand and normative impartiality on the other.[32]

The basic aim of this approach is ethical and practical (although Calarco himself points out that it often fails to be a successful ground for actual practice because of its overwhelming focus on theoretico-logical soundness, which leaves its propo-nents wallowing in endless debates rather than proposing institutional solutions), and views the human–animal distinction as what it seems to be at face value: a difference between one species of animal and others. The identity approach theo-rists claim that the difference has been artificially inflated into an absolute frontier, allowing for the mistreatment of animals which could be avoided if we learned to perceive a fundamental identity between man and (at least some) animals. While driven by indubitably noble intentions to form a theoretical framework allowing one to successfully counter animal cruelty and other forms of abuse, the pro-ponents of this approach fall short of the task they treat as fundamental to their endeavor – getting rid of the human–animal caesura.

Although the identity-based approach might be considered an umbrella-term for a vast field of refined and diverse standpoints, it is possible to reconstruct a bottom-line argument that will suffice to understand the point I am trying to make. Based on the underlying assumption that the interspecies differences are quanti-tative and not qualitative, it would go as follows: (1) the inclusion in the moral community (i.e., community of those who are at least moral patients) is based on having a certain trait/ability (reason, speaking etc.); (2) if this ability is found in a representative of species other than human, even if in a lesser degree, the representative (or the species) in question should be granted access to the moral community; (3) the traits in question are found in certain species (apes, dolphins, etc.), and therefore they should be included in the moral community.[33]

This tendency is visible, for example, in some aspects of Donna Haraway's thinking – even though Haraway herself would probably be critical of such an assessment. Analyzing Derrida's encounter with his cat, Haraway criticizes the French philosopher for not having recognized the true scope of possible interactions he might have had with the animal. She writes:

> But how much more promise is in the questions, Can animals play? Or work? [. . .] Can I, the philosopher, respond to an invitation or recognize it when it is offered? What if work and play [. . .] open up when the possibility of mutual response, without names, is taken seriously as an everyday practice available to philosophy and to science?[34]

I have already indicated that the category of response (and – *a fortiori* – responsibility) has been crucial for distinguishing man from animal for thinkers from Descartes to Cassirer, Lacan and beyond. Also, the notion of work (and, as far as connected with it, the notion of play)[35] has, at least since Marx, been used in this way.

Some pages later, analyzing the research of Barbara Smuts, who, much like Jane Goodall, spent a decent amount of time in ape societies – in Smuts's case they were baboons – Haraway uses another set of categories. She discusses how Smuts needed to shift from the then-widely accepted stance of an objective observer whose goal is to remain invisible to establishing a social relationship with the baboons, who were "unimpressed by her rock act."[36] Haraway comments:

> I imagine the baboons as seeing somebody off-category, not something, asking if that being were or were not educable to the standard of a polite guest. The monkeys, in short, inquired if the woman was as good a social subject as an ordinary baboon, with whom one could figure out how to carry on relationships, whether hostile, neutral, or friendly. The question was not, Are the baboons social subjects? but, Is the human being? Not, Do the baboons have "face"? but, Do people?[37]

Here, once more, Haraway's tactic is to present the animals as not only equal to humans, but as equally human (or even much more human than an "objective scientist"), as *zooi politikoi*, as "somebodies" and not "somethings," as beings endowed with a "face."

Finally, even the word she uses for the relationship that remains central to the discussed text is, if not chosen, then at least interpreted in a similar way. When dissecting the term "companion species," she notes "*Companion* comes from the latin *cum panis*, 'with bread'".[38] It is hard to imagine an object more traditionally associated with humanity than bread. It is also hard not to notice that establishing a companionship with animals by means of bread brings to mind a communion with a typically Western kind of deity.

By means of such rhetorical tactics, the "identity" approach not only fails to accomplish what it claims to do in the first place – i.e., work from a point where there is no radical human–animal caesura – but in fact reinforces the traditional boundaries, and, moreover, blurs them by pretending there is no caesura at all. It proves to be a fulfillment of the apparent logic of the man–animal difference: since we are special because we are rational (ethical, social etc.), then all that is rational (etc.) is special. Instead of effacing the caesura, it only displaces it, thus extending the human beyond man to include certain animals, and thus proves to be simply another installment of what Agamben would call the "anthropological machine," even though it pretends to do away with the terms in which the machine worked for centuries – man and animal. But the divisions it puts into place are based on the same logic that have formed the anthropological machine in the first place.

The discussion of the "identity" approach also reinforces the point made thanks to the reading of Derrida, namely that the only way to comprehend the roots of the radical difference between man and animal is to take the difference itself seriously. In order to understand the *logos* of philosophical zoo*logy*, one must first learn to live among those who mastered it. A just investigation can be completed only thanks to an approach that allows to combine two seemingly disparate traits found in those thinking in terms of the caesura: (1) the autonomy of philosophical thinking, which makes it immune to the supposed ontological implications of the sciences (especially evolutionary biology); and (2) the inclusive nature of philosophical thinking, which allows it to integrate or be inspired by the sciences, theology, the arts, etc. Therefore, I cannot continue before first obtaining at least a working definition of the specificity of philosophical thinking and the nature of argumentation which can be used to explore the problematic nature of philosophical zoology.

Notes

1 Matthew Calarco, "Identity: Difference: Indistinction," *CR: The New Centennial Review* 11, no. 2 (2011): 53–54.
2 Calarco, "Identity: Difference: Indistinction," 54.
3 This of course alludes to Derrida's exclamation "The animal, what a word!", in: Jacques Derrida, *The Animal That Therefore I Am*, trans. David Wills (New York: Fordham University Press, 2008), 23.
4 Derrida, *The Animal . . .*, 34.
5 I refer to man in the masculine not because of a lack of sensitivity to gender issues, but precisely because of this sensitivity, as I believe what is known as "man" has formed by a rejection of not only the animal, but also the woman, the child etc. I come back to this issue in Chapter 7.
6 René Descartes, *The Philosophical Writings of Descartes, Vol III: The Correspondence*, trans. John Cottingham, et al. (Cambridge: Cambridge University Press, 1991), 304.
7 Martin Heidegger, *The Fundamental Concepts of Metaphysics: World, Finitude, Solitude*, trans. William McNeill and Nicholas Walker (Bloomington and Indianapolis: Indiana University Press, 1995), 186.
8 Friedrich Nietzsche, *Thus Spoke Zarathustra*, trans. Adrian Del Caro (Cambridge: Cambridge University Press, 2006), 7.
9 Descartes, *The Correspondence*, 304.

10 Michel de Montaigne, *The Complete Essays*, trans. Michael Andrew Screech (London: Penguin, 2003), 280.
11 Carolus Linnaeus [Carl von Linné], *Menniskans Cousiner*, ed. Telemak Fredbärj (Uppsala: Ekenäs, 1955), 4–5; quoted in: Giorgio Agamben, *The Open: Man and Animal*, trans. Kevin Attell (Stanford: Stanford University Press, 2004), 24.
12 For a thorough analysis of this issue, see, e.g., James Rachels, *Created from Animals: The Moral Implications of Darwinism* (Oxford and New York: Oxford University Press, 1990), 1 and ff.
13 Ernst Cassirer, *An Essay on Man: An Introduction to a Philosophy of Human Culture* (New York: Doubleday, 1956), 44–61.
14 Ludwig Wittgenstein, *Tractatus Logico-Philosophicus*, trans. D. F. Pears and B. F. McGuinness (London and New York: Routledge & K. Paul, 1961), 49.
15 *Genesis* 1: 28.
16 In fact, Linzey seems to think that the attitude to animals that we consider typically Christian and was largely shaped by Aquinas comes more from the Aristotelian part of Aquinas' thinking than from the Christian part. See. Andrew Linzey, *Animal Theology* (Chicago: University of Illinois Press, 1994).
17 Diogenes Laertius, *Lives of Eminent Philosophers*, trans. Robert Drew Hicks (Cambridge, MA and London: Harvard University Press/William Heineman, 1956), vol. II, 43.
18 As Elizabeth de Fontenay remarks, because of the theory of metempsychosis, which allows souls to pass from man to animal and back again, the Platonic view of the man–animal frontier remains much more complicated than the tradition of regarding him as almost a proto-Christian idealist would like us to believe. On the other hand, Diogenes' gesture proves that he "remains caught up in the rising tradition of humanism at the very same moment when he believes to be dismantling it." See Elizabeth De Fontenay, *Le silence de bêtes* (Paris: Fayard, 1998), 156. In Plato, the truly important frontier lies not between man and animal, but between the free man and the slave; see Jean-Louis Poirier, "Éléments pour un zoologie philosophique," *Critique*, no. special, "L'Animalité" aout-septembre (1978): 673–88, quoted in De Fontenay, *Le silence . . .*, 156.
19 See, e.g., Gilbert Simondon, *Two Lessons on Animal and Man*, trans. Drew S. Burk (Minneapolis: Univocal, 2011), 50–53.
20 Leonard Lawlor, *This Is Not Sufficient: An Essay on Animality and Human Nature in Derrida* (New York: Columbia University Press, 2007), 24–26.
21 See, e.g., Jacques Derrida and Jean-Luc Nancy, "Eating Well," in *What Comes After the Subject?* ed. Eduardo Cadava, Peter Connor, and Jean-Luc Nancy (London and New York: Routledge, 1991), esp. 111–14.
22 Derrida, *The Animal . . .*, 30–31.
23 Jacques Derrida and Elizabeth Roudinesco, "Violence Against Animals," in *For What Tomorrow: A Dialogue*, trans. Jeff Fort (Stanford: Stanford University Press, 2004), 66.
24 Paola Cavalieri, "A Missed Opportunity: Humanism, Anti-humanism and the Animal Question," in *Animal Subjects*, ed. Jodey Castricano (Waterloo, Ontario: Wilfrid Laurier University Press, 2008), 106.
25 Cavalieri, "A Missed Opportunity," 111.
26 As Deleuze explains in *Nietzsche and Philosophy*, "[w]hen we ask what beauty is, we ask from what standpoint things appear beautiful: and something which does not appear beautiful to us, from what standpoint would it become so?". Gilles Deleuze, *Nietzsche and Philosophy*, trans. Hugh Tomlinson (London and New York: Continuum, 1993), 77. It is interesting to note is that rather than opting for the simple translation of the French *qui* ("who"), Hugh Tomlinson decided to render it as "which one").
27 Derrida, *The Animal . . .*, 62.
28 Derrida, *The Animal . . .*, 41.

29 Derrida, *The Animal . . .*, 9.
30 It might be interesting to note that Derrida never refers to the cat by her first name, as if such a gesture would cancel out the true individuality of the encounter. This seems to render problematic Leonard Lawlor's interpretation, who insist that "Derrida is even arguing, or at least this is what I am arguing: we must name all the animals with proper names, eliding all the definite articles"; Lawlor, *This Is Not Sufficient*, 104.
31 Derrida, *The Animal . . .*, 30.
32 Calarco, "Identity: Difference: Indistinction," 42.
33 A more detailed discussion of this argument, as well as the most complete analytical-philosophy account of theoretical animal studies, including a discussion of the "argument from marginal cases," which addresses an apparent flaw of this argumentation (that if we base moral status on rationality, language or other traits, and not species membership, then we should grant moral status to certain humans, who do not express these traits) can be found in: Paola Cavalieri, *The Animal Question: Why Nonhuman Animals Deserve Human Rights*, trans. Catherine Woollard (New York: Oxford University Press, 2001).
34 Donna Haraway, *When Species Meet* (Minneapolis: University of Minnesota Press, 2008), 22.
35 Surely, even philosophers know that animals play, but they traditionally do not participate in the work/play dialectics of the human.
36 Haraway, *When Species Meet*, 24.
37 Haraway, *When Species Meet*, 24.
38 Haraway, *When Species Meet*, 17.

2 Polemology

The animal as a philosophical concept and how that influences infra- and interdisciplinary discussion

As the previous chapter has shown, the most concise summary of philosophical zoology would be this: (1) within its limits, man and animal are essentially separated, so that a strict qualitative caesura exists between the two; (2) it is only understandable within the discipline of philosophy – it seems to defy common sense and logic as well as evolutionary biology. A corollary to the second point is that the reasons for the existence of the caesura can be understood only from within philosophical discourses that take it seriously, because – as I have already shown – no "outside" thought, be it biology, history or the critique of religion, can give justice to it. This statement seems to put philosophy (at least when it tackles the case of the animal) in an awkward position of a discipline which fails to take into account any outside source of influence – treated as "inspiration" to be contemplated or "truth" to be accepted – and remains confined in the autistic realm of its own musings. However, this too has been shown as untrue in the case at hand, with the examples of Descartes, Heidegger and Cassirer, whose works show a clear affinity with the scientific achievements of their respective times.

The two pillars of philosophical zoology thus turn into questions about philosophy as a discipline or way of thinking. The first one is internal – in what way is philosophy special, so that it allows for thinking in terms of the strict man–animal caesura? In the context of the case at hand, answering this question will help to provide a framework for the elaboration on the initial problem: what governs the difference between man and animal, how it is distributed and why it was so pervasive in the history of philosophy. The second question is external – what is the relationship between philosophy and other disciplines, especially science (in this case, biology), but also art, literature, cinema etc.? How is it possible that philosophy at once uses these disciplines profusely and seems to reject some of their most widely accepted theses (in this case, especially the thesis that there is only a quantitative, and not a qualitative – not to mention abyssal – difference between man and animal)? Is it a dialogue of two sides who cannot hear one another, a struggle for power, a case of serial misreadings or something else altogether?

In this chapter, I will try to show that – at least in the framework of the task at hand – the most useful way to think about philosophy can be found in the work of Deleuze (and Guattari).[1] Rather than reconstruct their theory of philosophy as a whole, I will focus on the most important points that help understand the investigation conducted in this book.

In a lecture preceding the 1991 book *What Is Philosophy?*, Deleuze makes a crucial – if at first glance trivial and controversial at once – statement: "No one needs philosophy to think."[2] Regardless of whether one needs philosophy at all, one does not need it for thinking; it is not the act of thinking that defines it. This understanding of philosophy opposes all of the currents stating that philosophy is a meta-discipline, which can be practiced in order to provide insight into other, "lower" disciplines, be they scientific or not. Deleuze:

> Treating philosophy as the power to "think about" seems to be giving it a great deal, but in fact takes everything away from it. [. . .] The only people capable of thinking effectively about cinema are the filmmakers and the film critics or those who love cinema. Those people don't need philosophy to think about film.[3]

The same is true for mathematics and mathematicians, painting and painters, history and historians, psychology and psychologists, poetry and poets, biology and biologists, etc.

The way of thinking Deleuze opposes is age-old, with its roots going as far as Aristotle. It received its perhaps most famous description in a metaphor Descartes included in a letter to Picot, the translator of *Principles of Philosophy*, serving as an introduction to the book:

> all Philosophy is like a tree, of which Metaphysics is the root, Physics is the trunk, and all the other sciences are branches that grow out of this trunk, which are reduced to three principal, namely Medicine, Mechanics and Ethics.[4]

Metaphysics is thus the First Philosophy, and the reasoning and effects of the other sciences rely on it being true and accurate. And though one might argue that in Descartes' times science and philosophy were still close enough to validate such a statement – for example, even Newton thought himself a philosopher and entitled his groundbreaking work *Mathematical Principles of Natural Philosophy* – but today the metaphor of the tree is anachronistic, we should remember that more than 300 years after Descartes, Martin Heidegger, in the "Introduction" to *What is Metaphysics?*, not only brings this vision back, but adds a comment that further reinforces it:

> Staying with [Descartes'] image, we ask: In what soil do the roots of philosophy take hold? Out of what ground do the roots, and thereby the whole tree, receive their nourishing juices and strength? What element, concealed in the ground and soil, enters and lives in the roots that support and nourish the tree?[5]

According to Heidegger, the character of philosophy is thus to ask not only for the roots of the tree of all science and knowledge, but further, to establish what soil nourishes the tree. This of course means, to put it in a Heideggerian language, to realize that metaphysics always already predetermines the answer to the properly philosophical question of what is "Being," and instead asks questions about beings. Heidegger sees philosophy – or, strictly speaking, fundamental ontology – as the truly grounding discipline, which sets a horizon for others, especially science.

This vision does not need to suggest such a dramatic (in a way anti-scientific) primacy of philosophy – the positivist thesis that philosophy should serve as "logical analysis of scientific language" is a more modest version of the same principle: philosophy's place is at a meta level, allowing it to assess science (or the sciences) from a point of view that is attainable only from that level.

As I already indicated, Deleuze's project is the exact opposite. It is at once humble – in the sense that philosophy does not claim the ability or indeed right to judge other disciplines – and distinctive – in the sense that it sees philosophy as a domain with its own identity, character and strengths. Let us first look how Deleuze introduces the notion of the "idea" (again, though the quotation pertains to cinema, it is as valid in regard to other disciplines):

> No one has an idea in general. An idea – like the one who has the idea – is already dedicated to a particular field. Sometimes it is an idea in painting, or an idea in a novel, or an idea in philosophy or an idea in science. And obviously the same person won't have all of those ideas. Ideas have to be treated as potentials already *engaged* in one mode of expression or another and inseparable from the mode of expression, such that I cannot say that I have an idea in general. Depending on the techniques I am familiar with, I can have an idea in a certain domain, an idea in cinema or an idea in philosophy.[6]

This means that regardless of the domain it expresses itself in, an idea always remains on the same level of expression; one cannot say, as Heidegger would, that in order to express, for example, a biological idea, one is always already constrained by what was established in another domain – in the case of biology this would mean the *philosophical* determination of what is life, the *bios* of biology.

Furthermore, there is no discipline that can talk about ideas *as such*, in a privileged manner; on the other hand, this potential entity enables cross-disciplinary understanding. For example, Deleuze notices a resemblance between Dostoevsky and Kurosawa, whose characters are

> constantly caught up in emergencies, and while they are caught up in these life-and-death emergencies, they know that there is a more urgent question – but they do not know what it is.[7]

Kurosawa expresses this idea in a cinematic manner, while for Dostoevsky it is a literary idea. However, while it cannot be understood outside of the given discipline, it still remains the same idea.

This is also why Deleuze is able to use examples taken from, or even write books about, areas different than philosophy – from cinema to Francis Bacon's paintings. One must, however, carefully take into account that under his (or any philosopher's) pen, they will always remain philosophical ideas.

Of course, Deleuze will refer to these "philosophical ideas" as concepts. While it is not my primary concern here to render in minute detail Deleuze's understanding of philosophy,[8] the concepts and the planes on which they are created, I will nonetheless need to point to certain characteristics of this understanding, which I will then use to sketch out a hopefully fruitful vision of how to sensibly analyze the question of the animal–human caesura within philosophy.

(1) **A concept is always formed of particles.** It is not a unity, but rather a haecceity, a fortuitous (or unfortunate) encounter of its components. In this way, the Platonic One is formed of being and nonbeing,[9] and the Cartesian *cogito* has three components: doubting, thinking and being.[10] Much as these components are inseparable within the concept, they themselves can be perceived as concepts – although this example is not given explicitly, it seems reasonably clear that Deleuze and Guattari would agree that both being and nonbeing are indeed concepts. It is thus natural that a concept is always linked to other concepts. In the words of Deleuze and Guattari:

> The concept is therefore both absolute and relative: it is relative to its own components, to other concepts, to the plane on which it is defined, and to the problems it is supposed to resolve; but it is absolute through the condensation it carries out, the site it occupies on the plane, and the conditions it assigns to the problem. As a whole it is absolute, but insofar as it is fragmentary it is relative.[11]

Whether we perceive a concept as absolute or relative is therefore always in itself relative to the perspective from which it is considered. This means that the fact that a concept can serve as a particle of another concept does not take anything away from its own potentially absolute nature; conversely, this absolute nature does not stop it from forming concepts or links to other concepts.

In such a vision, the difference between the whole and its parts – at least on the level of concepts – is multiplied and complicated, which will have important consequences for the understanding of the human–animal caesura. I will come back to this.

(2) **A concept has a history.**[12] This is partly an expansion upon the previous point, as the historicity of a concept is a special case of interconceptual connection. In particular, it means that a concept is linked to other, earlier concepts that bear the same name, but are somehow different – Deleuze and Guattari give the example of the Other as a possible world, which pertains to Leibniz as it does to Wittgenstein, albeit in a radically different manner. Tracing the history of a concept would therefore have to take into account the similarity allowing for a diachronic use of the same name for the concept, but also the differences between its incarnations – are they formed from different

particles? Do they express different ideas or answer to different problems or Events? (See the following point 3.) Are they created on different planes? (See point 4, to follow.)

(3) **The concept is at once non-referential[13] and it is always created in an answer to an Event.[14]** Thus, a concept refers only to itself, but it is connected to other concepts structurally and historically, and it answers a call of an Event which poses a problem. The true frontier lies not between the world and our conceptual representation of it, but between the particles of the concept, the relations it forms with other concepts, the historical relationship it forms with itself, and the Event to which it is an answer.

This last question is particularly interesting and seemingly paradoxical. It could appear that the Event is the exact equivalent of the Kantian *Ding an sich*, the Thing that needs to exist for us to have the material to form concepts about; the Lacanian Real around which the chain of the Symbolic is woven. The Event would thus be the beginning, the initial, *arché*-ic moment that begs for the concept to be formed. However, this cannot be the case, since concepts are non-referential – their connections are not formed on a vertical, but on a strictly horizontal, basis. This means that, much as in *Difference and Repetition* difference as the *arché* is always already a repetition,[15] the Event must already be a concept that shows itself in a different context (i.e., a different problem or a different plane).

But the association with Kantianism is not accidental. Deleuze's philosophy as concept creation is closely linked to his idea of transcendental empiricism. As Miguel de Beistegui explains:

It is [. . .] absolutely crucial not to model (*décalquer*) the transcendental after the empirical, as Kant does. Equally though, the transcendental must be explored for its own sake, and thus experimented with. Philosophy is concerned with experiencing, experimenting with, the transcendental itself, thus indicating the sort of work involved in creating concepts.[16]

A concept is thus a paradoxical, autoreferential lens[17] – on the one hand, conceptual creation occurs on the transcendental level, thus forming conditions of possibility for any philosophical experience; on the other hand, the lenses are created for their own sake, not to see something that supposedly lies "on the other side." In Deleuze's view there is no other side – at least not in this sense (see the following point 4) – in fact, there are no "sides" at all: there is only the plane, concepts and its particles.

(4) **Concepts are created on a plane.** According to *What Is Philosophy?*, concepts are created on what Deleuze and Guattari call a plane of immanence (they sometimes also use the name "plane of consistency"). Even though the same elements can form a concept and a plane of immanence (as it is in the already-cited case of doubting, thinking and being), the authors stress one needs to be wary not to confuse the two.

Concepts and the plane are as correlative as they are indivisible, and it is impossible to have one without the other. The concepts populate the plane, but at the same time they set it out. The plane dictates the rules for the creation of concepts, but at the same time the concepts, resonating with each other, create their plane. The division between the plane and the concept thus seems – at least for a philosopher's use – methodological or strategic. It is only in this sense that Deleuze and Guattari say that the plane is prephilosophical:

> philosophy posits as prephilosophical, or even as nonphilosophical, the power of a One-All like a moving desert that concepts come to populate. Prephilosophical does not mean something preexistent but rather something that *does not exist outside philosophy*, although philosophy presupposes it. These are its internal conditions.[18]

I will come back to the nonphilosophical nature of the plane next (point 5); here I will focus on the notion of the plane as such.

Actually, Deleuze differentiates two types of planes. To understand what the plane of immanence is "in itself," and not just in relation to the concepts that populate it, it is best to consult a different text of Deleuze's, first starting with the negative of the plane of immanence:

> On the one hand, a plane that could be called one of *organization* [. . .], [which] possesses a supplementary dimension, one dimension more, a hidden dimension, since it is not given for itself, but must always be concluded, inferred, induced on the basis of what it organizes. [. . .] It is therefore a plane of transcendence, a kind of design, in the mind of a man or in the mind of a god [. . .]. One such plane is that of the Law.[19]

An architect's design of a building is a perfect example of the plane of transcendence or organization, perhaps even on two levels – one is the set of rules the architect must observe in order for the building to serve its purpose (the laws of utility), to be appreciated (the laws of aesthetics), and simply not to fall apart (the laws of physics); the other is the conformity between the design and the building itself – the design serves as a transcendent model, which precedes the building in an ideal fashion.

> And then – says Deleuze – there is a completely different plane which does not deal with these things: the plane of Consistence [. . .] It is this plane [. . .] that might be opposed to the plane of organization. It is truly a plane of immanence because it possesses no dimension supplementary to what occurs on it; its dimensions grow and decrease with what occurs on it, without its planitude being endangered [. . .]. This is no longer a teleological plane, a design, but [. . .] an abstract drawing.[20]

The plane of immanence, as established by philosophy (or *a* philosophy), only obeys its own rules, and never subjects itself to anything that occurs outside.

Without the "outside" the plane of immanence cannot be immanent *to* anything; it is positively immanent, having or referring to no outside.

Francois Zourabichvili links the plane of immanence with Deleuze's theory of the univocity of being[21] (which he calls "without a doubt Deleuze's most important contribution to the history of philosophy").[22] This theory, inspired by Duns Scotus, and – at least according to Deleuze – dear to Spinoza and Nietzsche, can be summarized as saying that anything that is, is in the same way. There is no difference in the way things are, so none of them are in a way that is superior or inferior. This, of course (and against Alain Badiou) does not mean that Deleuze is dreaming of a great Unity or God under which (or whom) all the beings can be subsumed. On the contrary, as he says already in *Difference and Repetition*: "the essential in univocity is not that Being is said in a single and same sense, but that it is said, in a single and same sense, of all its individuating differences or intrinsic modalities."[23] While this early fragment pertains to all beings, in the special case of concepts, this means that no concept is, in virtue of its way of being on the plane, "better" or "privileged." On a plane of immanence, all their difference lies in the way they are intrinsically built and connected to other concepts.

However, if philosophy is indeed characterized as setting up a plane of strict immanence, its history – the history of "real" philosophy – is extremely short. It includes the Greeks, or, strictly speaking, only presocratics, because in the philosophy of Plato and his successors:

> instead of the plane of immanence constituting the One-All, immanence is immanent "to" the One, so that another One, this time transcendent, is superimposed on the one in which immanence is extended or to which it is attributed.[24]

A few centuries later,

> It gets worse with Christian philosophy. [. . .] all philosophers must prove that the dose of immanence they inject into the world and mind does not compromise the transcendence of a God to whom immanence must be attributed only secondarily.[25]

Descartes and his followers, including the phenomenologists, enclose immanence within the subject, which with Kant becomes transcendental – but the transcendental is just

> a modern way of saving transcendence: this is no longer the transcendence of a Something, or of a One higher than everything [. . .], but that of a Subject to which the field of immanence is only attributed by belonging to a self that necessarily represents such a subject to itself (reflection).[26]

And it does not stop at this point, for example, Jaspers' Encompassing only serves as "a reservoir for eruptions of transcendence."[27] It seems that most philosophy, for these or other reasons, was always already botched philosophy, which allowed or invited transcendence into its plane, and thus stopped being philosophy:

Whenever there is transcendence, vertical being, imperial State in the sky or on the earth, there is religion.[28]

Religion in the sense of transcendence (no wonder that "it gets worse with Christian philosophy"; on some pages Deleuze and Guattari even wonder, strangely echoing Heidegger, if Christian "philosophy" is philosophy at all) is philosophy's dark companion, and every time transcendence is introduced to a plane of immanence, philosophy becomes theological, and concepts become figures.

If one wished to speak strictly, there was only one philosopher (at least since the presocratics) – Spinoza. He is cited as the one who never let transcendence enter his plane; he is dubbed the "Prince" or even the "Christ" of philosophers. His plane of immanence was the "best."[29] There are some that would qualify at least partly: a few passages of Sartre, the British empiricists (one thinks of Hume especially), Nietzsche, Bergson (but only in the beginning of *Matter and Memory*), as well as some non-philosophers, such as Kleist and Hölderlin – it is easy to recognize the protagonists of Deleuze's "minor tradition" in philosophy.

The strict character of this differentiation is a point where I must stray from Deleuze's view, and this for two reasons. Firstly, since the need for this part of Deleuze's thought in this part of the text is strictly methodological, I cannot agree to such a cut within the tradition that I am about to analyze in the later part of this text, especially in the chapter "Anthropology." What is needed is a useful tool that could help to establish if there exists a common trait within this tradition, as it pertains to the human/animal frontier. Secondly, it does not seem to be Deleuze and Guattari's aim to dismiss such a significant part of the history of philosophy as non-philosophical: what they mean is that philosophy has an aim, and this aim is to create concepts on a plane of immanence – but for one or another sort of reason, most philosophers have not succeeded in this aim, thus degenerating into thinking on a plane of transcendence.

The difference between the plane of transcendence and the plane of immanence will be crucial in the next parts of his study, and deciding on using only one of them (or indeed the "strict" notions of philosophy and of the concept) would be begging the question. Thus, a more general term is needed to help establish a structure that would allow for such an impartial analysis.

Deleuze and Guattari provide such a term in *What Is Philosophy?* when they often follow the notion of the "plane of immanence" with that of the "image of thought" – sometimes they even omit the first one altogether. However, the notion of the image seems problematic, as it invokes visual themes I would not like to put forward and plays a problematic role when put in the context of Deleuze's philosophy as a whole (e.g., is it *the* or *an* image of thought?).[30] I will therefore use a term that cannot be found explicitly in Deleuze's own writings – even though it can be found in commentaries, notably in Francois Zourabichvili's

Deleuze: A Philosophy of the Event[31] – namely "plane of thought." This slightly unorthodox move forms a platform allowing to speak about the "prephilosophical" suppositions of a thought regardless of it being truly "philosophical" or rather "religious" (i.e., based on a plane of transcendence).

(5) **Concepts are a typically philosophical kind of ideas.** I have already touched upon this subject earlier, but it needs to be reiterated to show the way that it pertains to the question at hand.

As Anne Sauvagnargues comments, "Deleuze calls 'Ideas' complexes of sensation that are not reducible to discursive signification, but that stimulate thought."[32] This is why Deleuze insists that one can never express an idea *as such*, but that it is always expressed as a philosophical, scientific, cinematic, literary (etc.) idea. "Touched" or "inspired" by the same "complex of sensation," the philosopher, painter or filmmaker react in a radically different way, each not only in their own medium (as, for example, both philosophers and writers write), but also in a manner proper for their discipline or domain. The example of Dostoevsky and Kurosawa from *What is a creative act?* cited near the beginning of this chapter showed how the same idea can be expressed in different domains.

This is crucial, because up to this point Deleuze's vision of philosophy seems overly distinctive, absolutely disconnected from other domains such as science or art. However, this is only half of the story. Deleuze's own work proves that a philosopher can successfully take on works from other domains and see them for their philosophical content. The notion of the idea can be used to explain this possibility, this connection between domains that does not subject one of them to the other, and does not directly influence one or the other – i.e., translating a scientific idea ("functive," to use Deleuze and Guattari's term) into a philosophical one does not need to mean anything to a scientist and does not in itself claim to be a scientific intervention.

This way of thinking helps pave the way to answer one of the questions asked in the preceding chapter, concerning the strange pickiness of philosophers when it comes to scientific theories: why the flow of blood in the case of Descartes? Why von Uexkull and not Darwin in the case of Heidegger? These choices were never made (solely) according to the scientific merit of the theories, but rather the philosophical usefulness of the ideas in question. In the case of Descartes, the discovery of the flow of blood from the perspective of a philosopher is only significant so long as it helps establish a concept of man as a conglomerate of *res cogitans* and *res extensa*.

Moreover, though concepts exist as concepts only within philosophy, as ideas they themselves can travel through to other domains. A hint of how it happens has already been given previously, when it was mentioned that the plane of immanence is not only pre-philosophical, but also essentially non-philosophical. This is what allows non-philosophers the access to philosophy, even though, or even especially because, their thought is not of the conceptual kind:

> The nonphilosophical is perhaps closer to the heart of philosophy than philosophy itself, and that means that philosophy cannot be content to be

understood only philosophically or conceptually, but is addressed essentially to non-philosophers as well.[33]

This means that philosophy, through its conceptual creation, is relevant – and "essentially" so – to non-philosophers, shaping the domain of common opinion. Regardless of how it happens and if philosophy is privileged in this aspect – seeing how scientific and technological vocabulary or metaphors forged in poetry become parts of common language, one can be almost certain that it is not – this naturally changes the stakes of any conceptual work.

What does this understanding of philosophy mean for philosophical discussion in general and this study in particular? First of all, it provides an explanation for the apparent impermeability of philosophy to scientific reason. No scientific ideas can be straightforwardly implemented in philosophy (and *vice versa*), which means that criticizing a philosophy from the point of view of scientific discoveries is misguided and will ultimately be unsuccessful. This would mean that the flabbergasting reluctance on the part of many philosophers to adapt Darwinian ideas into their thought about the man–animal question is in fact simple – the ideas themselves fail to translate into adequate concepts or change the conceptual framework or the plane upon which the concepts are created. (A few cases of this tendency will be addressed in Chapter 5).

Secondly, the understanding of philosophy as creation changes the way that the *polemos* happens *within* philosophy. Concepts should not be judged "right" or "wrong" in themselves, or assessed as more or less correctly describing this or that "reality," but examined to see what particles they are formed with (or what larger concepts they are particles of), and what plane of thought they are created upon. It is philosophically useless to say that Heidegger was wrong to believe man is "world-forming" in contrast to the "worldless" animal, or that Descartes was wrong to say animals are machines who lack free will or the ability to properly feel pain. What, on the other hand, is useful, is to ask how the concepts of man and animal are connected, and whether they are created upon the plane of immanence or transcendence.

Deleuze's vision of philosophy and its relationship to other disciplines resembles the one presented by Jakob von Uexkull towards the ending of his *Foray into the Worlds of Animals and Men*, and it might be useful to quote the German ethologist here, especially that he was a very important influence for Deleuze, notably in his readings of Spinoza (as we will find out in Chapter 6). As von Uexkull says:

The environments of a researcher of airwaves and of a musicologist show [an] opposition. In one, there are only waves, in the other, only tones. Both are equally real. And on it goes in this way: In the behaviorist's environment of Nature, the body produces the mind, but, in the psychologist's world, the mind produces the body.

The role Nature plays as an object in the various environments of natural scientists is highly contradictory. If one wanted to sum up its objective characteristics, only chaos would result.[34]

While Deleuze's vision is different from von Uexküll's in many important points – for example, their understanding of "Nature," which for the self-described Kantian ethologist would be something else altogether than for the anti-Kantian transcendental empiricist Deleuze – this description seems to suit one as well as the other, at least on a general level. There is no point in asking which conceptual system is "right" and which is "wrong" – there are more crucial questions to be asked, such as whether one is more "interesting" than the other; if it points us to an important problem or Event; if it helps to frame it in a way that is more constructive. This vision helps to understand philosophy as a productive endeavor, meaning that rather than focus on argumentation and debate, what a philosopher should do is act by means of establishing planes and creating concepts.

These points help pave the way for the remainder of this work. In the next chapter, I will focus on understanding the concepts of man and animal (and the caesura between them), and their relationship, asking how they are formed, if they are part of a bigger structure, or if one is included in the other. Thus, I will perform an act of what could be called "philosophical anatomy." Having understood this relationship, I will be able to ask another question – what plane of thought are we operating on when thinking in terms of a radical man–animal caesura? I will then take a step back and assess an important drawback in how many supposedly non-anthropocentric philosophies create the concept of the animal. Finally, I will try to take a step forward and outline the paths along which the concept of the animal should be created anew.

Notes

1 Gilles Deleuze and Felix Guattari, *What Is Philosophy?* trans. Hugh Tomlinson and Graham Burchell (New York: Columbia University Press, 1994), the most important of the books that I will be referring to, was signed by both Deleuze and Guattari, however it is plausible to think that Guattari's impact in the stages of writing was minimal, and what the book owes him remains rather in the sphere of the concepts used (e.g., deterritorialization) and editing. See e.g., Francois Dosse, *Gilles Deleuze and Felix Guattari: Intersecting Lives*, trans. Deborah Glassman (New York: Columbia University Press, 2010), 456: "*What Is Philosophy?* was manifestly written by Deleuze alone, but he agreed to a coauthor credit with Guattari, as a tribute to their exceptionally intense friendship, suggesting too that the ideas developed in the book and its language were the fruit of their common endeavors since 1969." However, to be fair, I need to note that some commentators strongly disagree. For example Jérôme Rosanvallon points that we should look to (sadly, unspecified) "texts published by both [Deleuze and Guattari] in the 1980s, before this last collaboration [i.e., *What Is Philosophy?*], to see that such an interpretation of the book's genesis not only deforms the material reality, but also goes against the fundamental logic governing the [entire] oeuvre they produced together." Jérôme Rosanvallon, *Deleuze & Guattari à vitesse infinie: Volume 1, De la vitesse infinie de l'être . . .* (Paris: Ollendorff & Desseins, 2009), 19.

2 Gilles Deleuze, "What Is a Creative Act," in *Two Regimes of Madness*, trans. Ames Hodges and Mike Taormina (New York: Semiotext(e), 2006), 46.

3 Deleuze, "Creative Act," 46.

4 Réné Descartes, "Letter of the Author to the French Translator of the Principles of Philosophy Serving for a Preface," in *Principles of Philosophy*, trans. John Veitch (Whitefish: Kessinger, 2004), 4.

5 Martin Heidegger, "Introduction to 'What Is Metaphysics'," in *Pathmarks*, ed. William McNeill, trans. Walter Kaufmann (Cambridge: Cambridge University Press, 1998), 277.

6 Deleuze, "Creative Act," 45.

7 Deleuze, "Creative Act," 50.

8 The reader can find such an account, for example, in Rex Butler, *Deleuze and Guattari's 'What Is Philosophy'* (London and New York: Bloomsbury, 2016).

9 Deleuze and Guattari, *What Is Philosophy?* 29.

10 Deleuze and Guattari, *What Is Philosophy?* 24.

11 Deleuze and Guattari, *What Is Philosophy?* 21.

12 "In short, we say that every concept has a history, even though this history zigzags, though it passes, if need be, through other problems or onto different planes," Deleuze and Guattari, *What Is Philosophy?* 18.

13 "The concept is defined by its consistency, by its endoconsistency and exoconsistency, but has no reference: it is self-referential; it posits itself and its object at the same time as it is created." Deleuze and Guattari, *What Is Philosophy?* 22.

14 Deleuze and Guattari, *What Is Philosophy?* 16.

15 Gilles Deleuze, *Difference and Repetition*, trans. Paul Patton (New York: Columbia University Press, 1994), p. 129: "there is no true beginning in philosophy, or rather that the true philosophical beginning, Difference, is in-itself already Repetition."

16 Miguel de Beistegui, *Immanence: Deleuze and Philosophy* (Edinburgh: Edinburgh University Press, 2010), 6.

17 It is not by accident that I use the image of the lens. In the beginning of his second book on Spinoza, Deleuze invokes the same image, quoting Henry Miller: "You see, to me it seems that the artists, the scientists, the philosophers were grinding lenses. It's all a grand preparation for something that never comes off. Someday the lens is going to be perfect, and we are going to see clearly, see what a staggering, wonderful, beautiful world it is." [Gilles Deleuze, *Spinoza: Practical Philosophy*, trans. Robert Hurley (San Francisco: City Light Books, 1988), 14.] It would perhaps be interesting to see how these lenses correspond with those Spinoza (the "prince," "Christ" of philosophers; see following) used to grind.

18 Deleuze and Guattari, *What Is Philosophy?* 40–41.

19 Gilles Deleuze and Claire Parnet, *Dialogues*, trans. Hugh Tomlinson and Barbara Habberjam (New York: Columbia University Press, 1987), 91–92.

20 Deleuze and Parnet, *Dialogues*, 92–93.

21 E.g. Francois Zourabichvili, *Deleuze: A Philosophy of the Event Together with The Vocabulary of Deleuze*, trans. Kieran Aarons (Edinburgh: Edinburgh University Press, 2012), 184, 195, 196.

22 Zourabichvili, *Deleuze . . .*, 212.

23 Deleuze, *Difference and Repetition*, 36.

24 Deleuze and Guattari, *What Is Philosophy?* 44.

25 Deleuze and Guattari, *What Is Philosophy?* 45.

26 Deleuze and Guattari, *What Is Philosophy?* 46. In Chapter 4 ("Anthropology") I present a slightly different and more detailed account on how Kant's philosophy is formulated on a plane of transcendence.

27 Deleuze and Guattari, *What Is Philosophy?* 47.

28 Deleuze and Guattari, *What Is Philosophy?* 43.

29 Deleuze and Guattari, *What Is Philosophy?* 60; even Deleuze and Guattari put this word in quotation marks, as if they could not treat their exaggerated admiration with absolute seriousness.

30 The notion of an image of thought has a long and adventurous history in Deleuze's work. It is present in it ever since he began philosophizing on his own – it gives its name to the pivotal, middle chapter of *Difference and Repetition*. Discussing the problem of beginnings in philosophy, Deleuze notices that even the seemingly most

un-*arché*-ic philosophical thought rests upon presuppositions that it borrows from the so-called common sense. Yet this "common" sense is far from universal – there will always be people who fail to share these premises. In the example of Descartes, which Deleuze uses, the phrase that is to be without presuppositions is of course "I think therefore I am." Yet, even here, at least three things are presupposed – to understand this phrase (and we assume that everybody's understanding should be equal), one has to properly comprehend what "thinking," "being" and "I" mean. (One does not fail to notice that this is the same example Deleuze will use, 23 years later, in *What Is Philosophy?*). Anyone who dares to question these common meanings, is a fool and cannot think "right." Other presuppositions include the notion that "everybody desires the truth." Deleuze concludes: "Postulates in philosophy are not propositions the acceptance of which the philosopher demands; but, on the contrary, propositional themes which remain implicit and are understood in a pre-philosophical manner. In this sense, conceptual philosophical thought has as its implicit presupposition a pre-philosophical and natural Image of thought, borrowed from the pure element of common sense. [. . .] It is in terms of this image that everybody knows and is presumed to know what it means to think. Thereafter it matters little whether philosophy begins with the object or with the subject, with Being or with beings, as long as thought remains subject to this image which already prejudges everything [. . .] For this reason, we do not speak of this or that image of thought, variable according to the philosophy in question, but of a single Image in general which constitutes the subjective presupposition of philosophy as a whole." (Deleuze, *Difference and Repetition*, 131–32.)

In *Difference and Repetition* it is therefore not *an*, but *the* Image of thought Deleuze speaks about, and it is precisely this image that his project will be directed against. Its goal would be thought without image, and not a different image of thought. Deleuze dubs the thinkers that share the image under an umbrella term of "philosophy of representation," and it is clear that the pair image-representation is not chosen by accident. What is more, the image of thought, as linked to such problems as truth, representation, designation or recognition, is, if anything, an example of a plane of transcendence.

However, things look slightly otherwise in the earlier book on Proust, from which the term "image of thought" originates. Here, the stakes are set differently. Deleuze recognizes the "'philosophical' bearing of Proust's work: it vies with philosophy. Proust sets up an image of thought in opposition to that of philosophy. He attacks what is most essential in a classical philosophy of the rationalist type: the presuppositions of this philosophy. The philosopher readily presupposes that the mind as mind, the thinker as thinker, wants the truth, loves or desires the truth, naturally seeks the truth." [Gilles Deleuze, *Proust and Signs*, trans. Richard Howard (Minneapolis: University of Minnesota Press, 2000), 94.]

On the one hand, it is clearly the same image of thought that is being fought against – we recognize it as the "presuppositions" of "rational" thought – on the other, though, it might be changed, there is a possibility of another image. Yet, the new philosophical endeavor (valuing singularity over universality, unconscious causes over conscious will, etc.) would still share the characteristic of thought as having an image – thus, at least on some level, would still amount to a philosophy of representation.

The term itself, "image," has become of central interest to Deleuze in the 1980s, when he wrote two of his books on cinema: *Time-Image* and *Movement-Image*. Now, not belonging to philosophy of representation – or any other – it lost its original weight, but also lost its immediate connection with thought. It is not the place here to discuss the meanings of the two double terms. Suffice it to say that used in such a variety of ways, and not clearly defined in its constant changes, the notion of the "image" seems to be of very limited use. However, the second part of the notion, that of "thought" seems to be worth saving: if both the plane of immanence (image of thought in *What Is Philosophy?*) and the image of thought (plane of transcendence) have something in common, is that they are both pre-philosophical forms or presuppositions of thought.

31 Zourabichvili, *Deleuze . . .*, 53. It might be worth noting that Zourabichvili uses the notion as if "in passing," equating it immediately with the "image" of thought.

32 Anne Sauvagnargues, *Deleuze and Art*, trans. Samantha Bankston (London: Bloomsbury, 2005), 10.

33 Deleuze and Guattari, *What Is Philosophy?* 41.

34 Jakob von Uexküll, *A Foray into the Worlds of Animals and Men with A Theory of Meaning*, trans. Joseph D. O'Neil (Minneapolis: University of Minnesota Press, 2010), 135.

3   Anatomy
What is the concept of man made of?

At times, philosophical anatomy needs to seek the help of its medical counter-part. Indeed, in a key introductory moment of his reasoning in *The Open*, Giorgio Agamben invokes an Aristotle-inspired 18th-century anatomist and physiologist, Marie François Xavier Bichat, whose work he uses to show what he deems a unique, novel and pertinently important understanding of the frontier between the human and the animal. Bichat, in his *Recherches physiologiques sur la vie et la mort*, recognizes that in any higher organism (an animal or a man) there are two types of life: animal life and organic life. The first one, "defined by its relation to an external world,"[1] is roughly the equivalent of what Aristotle called by the same name, while the second one, "organic life" is a "habitual succession of assimila-tion and excretion"[2] – the equivalent of Aristotle's "nutritive life." Agamben then continues to invoke Bichat's metaphor:

> [I]t is as if two animals lived in every higher organism: *l'animal existant au-dedans* – whose life, which Bichat defines as "organic," is merely the repe-tition of, so to speak, blind and unconscious functions (the circulation of blood, respiration, assimilation, excretion etc.) – and *l'animal existant au-dehors* – whose life, for Bichat the only one that merits the name of "animal" is defined through its relation to the external world. In man, these two animals live together, but they do not coincide; the internal animal's {*animale-di-dentro*} organic life begins in the fetus before animal life does, and in aging and in the final death throes it survives the death of the external animal {*animale-di-fuori*}.[3]

For Agamben, at least three consequences of key importance stem from Bichat's distinction of the two modes of life. Firstly, the separation of nutritive and animal life allows for practices of modern medicine – anesthesia, put in simple terms, is "turning off" the animal life while the vegetative remains "on." Also, modern biopolitics rest on the premise that it is vegetative life that needs to be taken care of and controlled.

Secondly, as Agamben stresses,

> [t]he division of life into vegetative and relational, organic and animal, ani-mal and human, therefore passes as a mobile border within the living man,

and without this intimate caesura the very decision of what is human and what is not would probably not be possible.[4]

If Agamben is right, then it is precisely man that one should focus on when considering the human/animal distinction. Man is, in a sense, a paradoxical being – if he consists of both human and animal, he is both a whole and a part of the whole, he is self-same and at the same time possesses an animal other within himself. It is only later that the full consequences of this consideration will become clear.

Thirdly, Agamben claims that the separation of nutritive life calls for a rethinking of the way we define the human in comparison to the animal:

> if the caesura between the human and the animal passes first of all within man, then it is the very question of man – and of "humanism" – that must be posed in a new way. In our culture, man has always been thought of as an articulation and conjunction of a body and a soul, of a living thing and a logos, of natural (or animal) element and a supernatural or social or divine element. We must learn instead to think of man as what results from the incongruity of these two elements, and investigate not the metaphysical mystery of conjunction, but rather the practical and political mystery of separation.[5]

This redefinition, thinking of man as the place of a mobile caesura, rather than a mysterious union, is believed by Agamben to have grave practical consequences, in the sphere of politics (including "great issues" such as the question of human rights) and our possible relationship to the divine.[6] The frontier moves from one place to another to form the different versions of answering the question "what is man"; the forces (political or religious) moving the frontier reinvent or reconstruct man each time. Agamben calls this process of establishing the frontier (and thus constructing man) the *anthropological machine.*

To put it in terms introduced in the previous chapter, what Agamben accomplishes with his act of philosophical anatomy is to sketch the relationship between three concepts. The concept "man" is created using two particles: "human" and "animal," and the relationship between these two particles, understood as separation, is changeable and determines what the two particles (which themselves can be understood as concepts) mean. What is more, one of these concepts – namely "human" – somehow dominates over the other. Before I can provide a more accurate picture of this domination and explore the nature of the relationship between the two concepts, I have to further assess Agamben's thesis.

Firstly, let me address a potential omission of Agamben's. Hardly anyone even vaguely familiar with Aristotle will fail to notice that Bichat is not distinguishing between human and animal (life), but between animal and plant (life). Agamben acknowledges this at first, only to seemingly forget about it when he equates the border between "vegetal and relational, organic and animal," and then, after another comma suggesting the same type of synonymous repetition he used before, "animal and human."[7] Up to this point he showed that, according to Bichat's reasoning, every higher living being is composed of "plant" life and "animal" life.

The only caesura we can truly speak of is the one between the vegetal/vegetative and the animal that runs within every living animal, including man. This is enough to draw the first of the three consequences listed previously: indeed, anesthesia is based on a temporary "shutdown" of animal life – the one that is capable of feeling pain – while leaving the vegetative core untouched. However, it makes the other two points Agamben tries to make – about the existence of a mobile caesura between animal and human inside man and the separation of animal and human within man as the defining act of any "humanism" – questionable.

And yet it would be imprudent to reject Agamben's thesis as altogether false. Even though the next few chapters are focused on showing the mobile character of the frontier, and not further pondering its character, at least one of them also hints to the missing link of the former argument. A few pages after formulating his thesis, Agamben analyzes the Thomistic "physiology of the blessed" or the state of the body of the resurrected residing in paradise. The argument and the problem are based on the separation between the human and the animal. First, Agamben quotes the following passage from the *Summa Theologiae*:

> those Natural operations which are arranged for the purpose of either achieving or preserving the primary perfection of human nature will not exist in the resurrection [. . .] and since to eat, drink, sleep and beget pertain to [. . .] the primary perfection of nature, such things will not exist in the resurrection.[8]

Agamben then comments that Aquinas

> proclaims unreservedly that animal life is excluded from Paradise, that blessed life is in no case animal life. Consequently, even plants and animals will not find a place in Paradise: "they will corrupt both in their whole and in their parts."[9]

One might argue that Agamben's quote from the *Summa* does not address animal life as described previously, but precisely the vegetative functions as present in animals (i.e., animal forms of assimilation, excretion and the somewhat more problematic function of reproduction). On the other hand, Aquinas does not speak of "human" life but of "blessed" life. That, however, does not mean that locating human essence in the capacity for being resurrected (as opposed to the animal/plant as the lack of such capacity) is incorrect. One should ask what this capacity is based on – the answer to this question will also show why Agamben's path does not tackle it directly, and why this line of questioning could blur some of Agamben's results.

Aquinas puts the case quite bluntly in a fragment of the very same question of the *Summa* that Agamben takes his citations from:

> Further, if the end [here in the meaning of *telos*, KS] cease, those things which are directed to the end should cease. Now animals and plants were made for the upkeep of human life; wherefore it is written (Gen. ix. 3) Even

as the green herbs have I delivered all flesh to you. Therefore where man's animal life ceases, animals and plants should cease. But after this renewal animal life will cease in man. Therefore neither plants or animals ought to remain.[10]

This argument, roughly conforming to the means/ends citation of Agamben's, is clear enough: as only means to an end (which is always man), animals and plants will disappear when not needed to sustain man. What is more, Aquinas explicitly states that "animal life" in man will cease as well. The boundary clearly goes between man and animal/plant and not plant and man/animal, as in Agamben's citation, and there is strong evidence of frontier between truly human life and animal life within man – the latter also being just a means of sustaining the former.

Whatever the reasons for Agamben not citing this fragment, it is certain that such a reading of Aquinas changes the stakes of the analysis of the human–animal frontier. A few lines afterwards, the *Summa* reads:

[W]hile men are corruptible both in whole and in part [contrary to incorruptible heavenly bodies, and elements, corruptible in part but not as a whole – K.S.], but this is on the part of their matter not on the part of their form, the RATIONAL SOUL to wit, which will remain incorrupt after the corruption of man. On the other hand, dumb animals, plants, and minerals, and all mixed bodies, are corruptible both on the part of the matter which loses its form, and on the part of their form which does not remain actually; and thus they are in no way subjects to incorruption.[11]

It is quite clear that the division between those who remain after the renewal of the world and those who (or: that which) perish(es) is based on whether the beings in question possess a rational soul. In an earlier-cited fragment, Agamben stressed that his idea of an intra-man division between the animal and the human goes against traditional definitions. But it seems clear that in Thomas' logic of resurrection, the difference between man and animal is based on a caesura that goes inside man (between the incorruptible rational soul and the corruptible animal parts). This puts into question Agamben's thesis that traditional ways of defining man are based on conjunction – as the logic of resurrection entails, being-man is being human (rational) and animal. The separation is already present, even though it is only actualized along with the fulfillment of the human essence of man in resurrection. In other words, the creation of the concepts "human" and "animal" is a specific separation inside the concept "man." What is more, the division is most clearly visible in definitions of man, treating him, for example, as a "rational animal," where the humanity clearly rests in the first part of the definition, excluding animality altogether.

This way of defining man is present much earlier in history and the symptoms of its logic can be found in Aristotle as well as in an earlier rejection of a famous Platonic definition. At first glance, the Platonic and Aristotelian (scholastic) definitions of man have the same classic structure – a genus is first given, and then a *differentia*, which shows how to clearly distinguish the species man from other representatives of this genus. Plato's "shorter way" of defining man in the *Statesman*[12] is simple to the extreme and surprisingly empirical for this philosopher:

> one must straight off at the start distribute the pedestrial by the two-foot in relation to the four-foot genus and – with the sighting of the human, that it still shares its lot with only the winged – cut the two-footed herd again by the stripped and wing-growing [or: feathered – KS] difference.[13]

Man is therefore a "featherless biped." However, one must notice what the specificity of this definition comes from. Neither featherlessness (which man shares with all animals save birds) nor the fact of walking on two feet (shared with birds but no other animals) are traits that produce an abyssal gap between animals and humans. What is more, the definition itself has a provisional nature[14] – one can easily imagine a different path, starting from featherlessness: first we divide animals into feathered and unfeathered, then into bipeds and quadrupeds, and we arrive at the same conclusion, even though the *genus proximum* and the *differentia specifica* are reversed.

This provisional and characteristically empirical (even naively so) character of this definition is perhaps best visible in the famous story about Diogenes of Synope I already mentioned in Chapter 1. Laertius recounts:

> Plato had defined Man as an animal, biped and featherless, and was applauded. Diogenes plucked a fowl and brought it into the lecture-room with the words: "Here is Plato's man." In consequence of which there was added to the definition, "having broad nails."[15]

What Diogenes sought was to ridicule the Platonic definition, showing that such a naively empirical way of defining man was doomed to failure. However, what his gesture did reveal about the Academy's way of defining man was its minimalistic character: it was based on a simple enumeration of easily recognizable, common animal traits of which man happens to have a unique combination. The addition of broad nails also fits within this schema (think of elephants). The only way such an addition can be made so easily, is when the definition itself is based on an intimate proximity between man and animal – something which is clearly (and somewhat paradoxically, given his dog-like – or *kynic* – nature) absent from Diogenes' thinking.[16]

What Aristotle opens up and what Aquinas and later philosophers exploit to form what one might call "traditional anthropology" is a different definition of man altogether. As I have already noted, following and diverting from Agamben, the "*rationale*" in *animal rationale* produces a single-cut caesura between what is properly human and what is animal, a dark, ultimately (in the logic of resurrection) useless particle. Its radicalism disallows a simple, Diogeneseque falsification, and the ultimate intangibility of what it refers to (this is not only a trait of rationality, but also of speech and living in a political way; I will have the opportunity to comment on this later). One does not need Descartes to know how difficult it is to prove that a man on the street is a rational being.

It is therefore crucial to stop reading traditional "Aristotelian" definitions of man as simple *genus-proximum-et-differentia-specifica*, aiming at providing a specific trait that will differentiate man from other animals, while still keeping him "inside" the animal kingdom. In fact, they provide an intra-man human trait that resides outside the reach of any animal, including the one that is inside man. As Thierry Gontier puts it, in the classic definitions formed according to the schema "man = X animal,"

[t]he "X" (reasonable, for example) is always understood [. . .] as both something that accomplishes animality in the highest degree and something that excludes man from this animality.[17]

Man might be considered the highest among the animals, but what the very gesture of defining in this classical way signifies is the exclusion of his very essence – what Gontier calls the "X" and what is actually the only trait we can call properly "human" – from animality, even the one inside man. To read these definitions correctly one would have to write them differently: man is not a rational animal, but is rational *and* animal, political *and* animal, ethical *and* animal – man is human *and* animal. He is not human so far as he is animal, and is not animal so far as he is human.

With Aristotle and those who followed him in this respect, the manner of creating the concept man has thus changed drastically. In Plato's case, the particles of the concept were chosen to support a practical scheme, one that has predominately a political end; they could be exchanged or reformulated if that end is not met. In the case of the scholastic definitions, the concept of man is created by means of an internal separation of "human" and "animal." The possibilities of shifts and changes between the particles of the concept are thus heavily limited, with one – namely "animal" – always staying in its place, regardless of the internal changes that it itself might undergo.

However, at least one possible breach in this tradition must be considered – the one that apparently happens with Descartes, who decided to radically change the way we define man, openly rejecting, along with the better part of the scholastic tradition, the definition of man as a rational animal.[18] This rejection merits a closer look. Let us join Descartes in the course of his second meditation, right after the strategically crucial juncture at which he had undoubtedly proven the existence of his doubting self.

What then did I formerly believe myself to be? Undoubtedly I believed myself to be a man. But what is a man? Shall I say a reasonable animal? Certainly not; for then I should have to inquire what an animal is, and what is reasonable; and thus from a single question I should insensibly fall into an infinitude of others more difficult.[19]

This fragment shows that Descartes' rejection of defining man as a rational animal has two linked, yet distinct, moments. What is logically first is his rejection of a way of defining one undefined concept with the use of other undefined ones, as this leads to a potentially infinite line of concepts that would take him nowhere close to what he sought to understand in the first place. Only then does he reject the concept of man – but only provisionally – and seeks to understand who he (i.e., Descartes or in any case the narrator of the *Meditations*) is. He starts by analyzing various traits, first belonging to what he used to understand as his body, then of what he had been taught to call the soul, and finally comes to the well-known conclusion:

> I am [. . .] a real thing and really exist; but what thing? I have answered: a thing which thinks.[20]

and by thinking he understands all of the possibilities pertaining to the soul:

> What is a thing which thinks? It is a thing which doubts, understands, [conceives], affirms, denies, wills, refuses, which also imagines and feels.[21]

"And what more?" – asks Descartes. And answers immediately:

> I am not a collection of members which we call the human body; I am not a subtle air distributed through these members, I am not a wind, a fire, a vapour, a breath, nor anything which I imagine or conceive, because I have assumed that all these were nothing.[22]

The last part of the sentence seems to slightly modify the content of the previous result, clearly hinting at regaining or reconquering his body by Descartes, which will happen in the course of the next meditation. But in fact, it does not alter the key point made earlier – a man is a thinking thing, a soul. We know a man has a body, it is true, but even if he were not to have it, he would be man nonetheless. The soul is crucial, essential, it is a necessary and sufficient condition for being a man, the body is accidental, it is unnecessary.

An animal, on the other hand, is pure body. For Descartes, all behavior of animals can be explained by mechanical movement, reactions to outside and inside stimuli – as a matter of fact, so can our "involuntary" movements,[23] such as sleep-walking or covering our faces when we fall. In fact, there would be no external difference between us and animals, save for one detail: speech, and a special kind at that:

> [S]poken words or other signs that have reference to particular topics without expressing any passion [. . .] I say that signs must have reference, to exclude the speech of parrots, without excluding the speech of madmen, which has reference to particular topics even though it does not follow reason. I add also that these words or signs must not express any passion, to rule out not

only cries of joy or sadness and the like, but also whatever can be taught by training to animals [because it involves inducing behavior by passion, say, for eating] [. . .] there has never been known an animal so perfect as to use a sign to make other animals understand something which bore no relation to its passions; and there is no human so imperfect as not to do so.[24]

I quote this part in its entirety, because it introduces one of the key themes of the human–animal distinction, namely the ability to respond and not just react, which is as important for Derrida as it is to Cassirer. To put it simply, an animal's reaction is always automatic, a man's response takes place regardless of, or even contrary to, his passions. This paves way for another specifically "human" trait, namely deceit.[25]

In other words, animals are like clocks or other machines hence their perfection. Animals do not make mistakes – they work perfectly according to the natural construction of the mechanisms which they are. So do bodies of men if they are somehow detached from their souls, for example as in the cases of sleepwalking or involuntary movements.[26]

What this means that man is indistinguishable from the animal but for the soul. The soul is the human in man, the only essential part of him, while the body is the animal in man, an accidental, mechanic particle. Hence the division between the human and the body-animal is strict and final.

A possible objection might be that Descartes is preoccupied precisely by the "conjunction of a body and a soul" that Agamben deems a badly conceived view of man, which he opposes to separation. Is not the problem of how the body and the soul interact the most enduring heritage of Descartes? Is not the absurd solution of the pineal gland the ultimate indication that here he approached the problem-of-problems?

In fact, an argument can be made that the opposite is the case. The very reason Descartes is unable to bridge the gap between the body and the soul is that he was so ingenious at separating them in the first place. According to this reading, the first two *Meditations* are not just an epistemological narrative, but also a meticulous process of distilling the human from within man, resulting in the creation of a new concept of the soul. The search for the indubitable starting point has as its effect – and as its stakes – the very division of the human and the animal, both within man and outside him.

Read in this way, the Cartesian rejection of the classical ways of defining man in fact reinforces the earlier-identified tendency of rejecting the animal from what is truly human in man. The soul has all the indications of the "X," a human trait raising man above animality, and Descartes' "definition" does not even try very hard to hide this. If Aquinas' man was animal *and* rational, Descartes' man is animal *and* soul.

The detour through scholastic and Cartesian definitions of man reinforces Agamben's thesis systematically (rather than by multiplying examples, of which *The Open* already contains a copious amount), allowing for the confirmation of the initial thesis of what was dubbed "philosophical anatomy" – the concept "man" is indeed formed of the particles "human" and "animal," which are separated by an ever-moving caesura. The caesura works in a paradoxical way – while the human is the essence of man, a necessary and sufficient for being one, the animal is always defined as what remains, a remnant of non-man in man. I will return to this problem later.

An additional effect of this investigation was the realization that definitions of man – especially the classical ones, like *animal rationale* or *zoon politikon* – are useful tools for locating the caesura. Indeed, their proposed reading, putting the particle "and" between the supposed *genus proximum* and *differentia specifica*, in fact lists the particles of the concept "man," thus pointing towards the path for further considerations.

This path strays away from Agamben, whose analysis has different stakes than mine,[27] and consists in answering the second of the questions signaled in the preceding chapter, namely what the plane of thought that the concept "man" is based on. I will argue that this question is best tackled by means of analyzing how classical definitions of man work in different (often very "unclassical") contexts.

Notes

1   Giorgio Agamben, *The Open: Man and Animal*, trans. Kevin Attell (Stanford: Stanford University Press, 2004), 14.
2   Agamben, *The Open*, 14.
3   Agamben, *The Open*, 14–15.
4   Agamben, *The Open*, 15.
5   Agamben, *The Open*, 16.
6   As we have seen before, the divine – or, at the least, transcendence – is indeed ever-present whenever we speak of the animal/human distinction.
7   Agamben, *The Open*, 15.
8   Agamben, *The Open*, 19.
9   Agamben, *The Open*, 19.
10 Thomas Aquinas, *The "Summa Theologica,"* trans. The Fathers of the English Dominican Province (London: Burno Oates & Washbourne, 1942), part III (supplement), p. 67.
11 Aquinas, *Summa Theologica*, 68.
12 One should not fail to notice that here also the definition of man is given in a political context, as in the opening paragraphs of Aristotle's, *Politics*.
13 Plato, "The Statesman," in *The Being of the Beautiful*, trans. and ed. Seth Bernadete (Chicago and London: University of Chicago Press, 1984), 226e. I keep the original translation ("wing-growing"), but in the light of Diogenes Laertius' story which I recount a few lines later, it seems that "feathered" better suits how Plato's definition was understood. If what the definition meant was indeed "wing-growing," then plucking the fowl (and not tearing its wings off entirely) would have been a pointless gesture.
14 See, e.g., Melissa S. Lane, *Method and Politics in Plato's Statesman* (Cambridge: Cambridge University Press, 1998), 42.
15 Diogenes Laertius, *Lives of Eminent Philosophers*, trans. Robert Drew Hicks (Cambridge, MA and London: Harvard University Press/William Heineman, 1956), vol. II, 43.
16 What might be especially interesting today is that this view opens up a radically different way of looking at man in his relationship to the "living" animals – a featherless,

broad-nailed biped belongs equally to the realm of the featherless, the realm of the broad-nailed and the realm of the two-footed. Crisscrossing genres, genus and species, man enters into surprising connections and disconnections (and so do other beings) – a similar view will be held in the last chapter of this work, although it will be based on the work of Spinoza rather than Plato.

17 Thierry Gontier, *De l'homme à l'animal: Montaigne et Descartes ou paradoxes de la philosophie moderne sur la nature des animaux* (Paris: Vrin, 1998), 16.

18 There has been a debate as to the extent to which Descartes' animal philosophy is a continuation of Aquinas' thoughts on the matter. E.g., Andrew Linzey and Peter Singer claim it is, while J. Barad opposes. However, this debate is oriented toward the evaluation of the value of animal life in both discourses (i.e., their ethical consequences) rather than the divide in question. See, e.g., Andrew Linzey, "Christianity and the Rights of Animals," *The Animals' Voice* 2, no. 4 (August 1989); Peter Singer, "Animals and the Value of Life," in *Matter of Life and Death*, ed. Tom Regan (New York: Random House, 1980); Judith Barad, *Aquinas on the Nature and Treatment of Animals* (London and San Francisco: International Scholars Publications, 1995).

19 René Descartes, "Meditations on First Philosophy," in *The Philosophical Works of Descartes*, trans. Elizabeth S. Haldane and G. R. T. Ross (Cambridge: Cambridge University Press, 1967), 150.

20 Descartes, "Meditations on First Philosophy," 152.

21 Descartes, "Meditations on First Philosophy," 153.

22 Descartes, "Meditations on First Philosophy," 152.

23 Since one of the soul's key functions is willing, "involuntary" easily translates into "soulless," which then can be understood as "inhuman," and finally "animal".

24 René Descartes, *The Philosophical Writings of Descartes, Vol III: The Correspondence*, trans. John Cottingham, et al. (Cambridge: Cambridge University Press, 1991), 303.

25 I only brush upon this notion, as it has been already intensely analyzed, e.g., in Lacan's notion of the "feigned feint" [see Jacques Lacan, *Écrits: A Selection*, trans. A. Sheridan (New York: Norton, 1977), 305] and its critique by Derrida (*The Animal That Therefore I Am*, trans. David Wills (New York: Fordham University Press, 2008), 119–40).

26 See Descartes, *The Correspondence*, 304.

27 Agamben seeks a way to destroy the caesura itself by means of a strategy aimed at rendering the anthropological machine inoperative, which amounts to the appropriation of the animal-within-man, the acknowledging of the hiatus of the intra-man division and exploring it. As Agamben puts it: "To render inoperative the machine that governs the conception of man will [. . .] mean no longer to seek new – more effective or more authentic – articulations, but rather to show the central emptiness, the hiatus that – within man – separates man and animal, and to risk ourselves in this emptiness: the suspension of the suspension, the Sabbath of both animal and man." (*The Open*, 92). His solution has been widely discussed and engaged critically from different perspectives: Heideggerian [e.g., Krzysztof Ziarek, "After Humanism: Heidegger and Agamben," *South Atlantic Quarterly* 107, no. 1 (Winter 2008): 187–209), feminist [e.g., Ewa Plonowska Ziarek, "Bare Life on Strike: Notes on the Biopolitics of Race and Gender," *South Atlantic Quarterly* 107, no. 1 (Winter 2008): 89–105], and from the point of view of its use for animal studies [e.g., Kelly Oliver, *Animal Lessons: How They Teach Us to Be Human* (New York: Columbia University Press, 2009), esp. chapter 10; Matthew Calarco, *Zoographies: The Question of the Animal from Heidegger to Derrida* (New York: Columbia University Press, 2008), chapter 3]. Both the scope of this work and the purely pragmatic character of the engagement with Agamben herein render it impossible to enter into in-depth conversation with these views.

4 Anthropology

Several instances of how the
concept of the human has
been established on a plane
of transcendence

The preceding part of this study helped set out the stakes for the analyses to follow. The human–animal frontier is a movable caesura that passes first and foremost within man. The two terms, human and animal, can be, in the context of a radical divide, treated seriously only as codependent philosophical concepts. Given the persistence of thinking in terms of a radical frontier despite the different loci it was situated in (between body and soul, between rationality and irrationality, will and automatism, politics and herd-life), it is safe to say that there is a deeper logic, a set of presuppositions, a pre-philosophical plane of thought, that is at work here forcing to think these two concepts in this way. According to Deleuze and Guattari's terminology introduced in Chapter 2, it could be a plane of transcendence or immanence.

The analyses conducted in this chapter will focus on several thinkers who treat the human–animal divide seriously, in most cases despite the fact that their philosophies were established after Darwin's discoveries. Each part of this chapter is devoted to a variation of the classical definitions of man, seeing him as distinct from animals based on language, politics, rationality and ethics. Of course, some of the thinkers do not just analyze one of the classical definitions – Heidegger will point out that *animal rationale* is a (mis)translation of the Greek *zoon logon ekhon* and work on both rationality and language, while in Kant's case rationality is closely linked to ethics. However, this impurity of the separate parts does not spoil the general purpose of the chapter, which is to see the fundamental logic behind each of these definitions – the fundamental way of thinking the difference between the human and the animal that allows all these definitions to arise.

Heidegger and the *zoon logon ekhon*

Few – if any – philosophers have been as widely analyzed and criticized by animal studies scholars as Martin Heidegger.[1] It would thus be pointless to provide a full-scope presentation of his take on the animal question. However, Heidegger's theory is crucial for the investigation undertaken here, namely, finding the plane of thought on which the concepts of man and animal are created. I will therefore proceed by sketching out a route through the German philosopher's thinking that – as I believe – best serves this purpose. If this analysis has a stand-alone merit outside of the immediate scope of this work, it lies in pointing out the intimate connection between crucial notions (animal, language, transcendence and death) that have usually been considered in separate pairs.

There are several reasons why Heidegger is so well suited to serve this book's purpose. Firstly, he was a post-Darwinian philosopher – and an avid reader of Nietzsche – and I have already alluded to his paying close attention to the sciences of his time, especially biology, in Chapter 1; his rejection of "biological continuism" implied by the theory of evolution must therefore come from a place other than simple ignorance.[2] The rejection of classically understood metaphysics is an important part of Heidegger's pursuit of a non onto-theological way of thinking, which makes his adherence to any type of "metaphysical separationism" at least problematic. And yet, Heidegger's thinking includes one of the strictest versions of the human–animal caesura in history, based on a particular reading of classical definitions of man.

Although Heidegger famously rejects (in *Letter on "Humanism"*) the humanistic or metaphysical definition of man as *animal rationale*, he does not claim that it is false. His other treatments of the definition (such as the one in *What is Called Thinking*)[3] suggest that this rejection comes not from the claim that a man cannot, or an animal can, think, but from the claim that it is a *humanistic or metaphysical* definition. Saying that it is metaphysical means that the definition presupposes something that Heidegger, ever since *Being and Time*, has been putting into question: a predetermination of an answer to the question of being. Man as a *rational* animal in the sense of metaphysics will always fail to ask the question of being. Man as a rational *animal* in the sense of metaphysics will always remain essentially an animal, even if we grant him or her the faculty of rationality as a *differentia specifica*. In Heidegger's own words: "Metaphysics thinks of the human being on the basis of *animalitas* and does not think in the direction of his *humanitas*."[4]

Heidegger thus understands man as a being *essentially* different from the animal and a being which (as we will later see) is essentially capable of asking the question of being. It is not by accident that I stress the word essential. It signifies that the question of man (and the human–animal divide) can only be answered, according to Heidegger, on the level of philosophy.

We have already seen in Chapter 1 that from Heidegger's standpoint, the difference between man and animal works by separating the beings in question into two (ontologically) equal classes, and that the class of "animals" is (again – ontologically) homogenous regardless of the size, species membership or capabilities of the given individual. In Heidegger's own words:

> It is [. . .] a fundamental mistake to suppose that amoebae or infusoria are more imperfect or incomplete animals than elephants or apes.[5]

This difference is best illustrated by another classical definition of man, that is *zoon logon ekhon*. Men differ from animals because they are essentially beings that have language. It is well known that *animal rationale* is a historical, Latin translation of this Greek phrase. However, in Heidegger's view, this is not simply a translation, but also an interpretation. It is in the interpretation of *zoon logon ekhon* that the essence of man has been presupposed as that of *animalitas* with an added trait of "rationality." Rather than follow this interpretation, Heidegger will claim that language is an essentially human trait that the animal lacks, therefore – to simplify the matter slightly – starting from the *humanitas* provided by language.

One must keep in mind that Heidegger understands language in a highly idio-syncratic way – not as a logic or a grammar, not as communication, but essentially as a certain *way of being*. That the human "has" language is only an approximate way of speaking; it would be truer to say that human beings *are* in the way of (having) language. Language is prior to thought, thinking is governed by language. This is why Heidegger can agree with a version of the definition of man as *animal rationale* – considered as a non-metaphysical statement, it means that human is essentially capable of thinking, and that means that he is capable of language.

As with the question of animality in general, much has been written about Heidegger and language,[6] and I cannot hope to address this question fully. I will limit myself to remarks that are directly linked to the initial question of this chapter, and only describe how Heidegger's understanding of language influences his distinction between man and animal.

Language is a way of being that is essentially human and not animal. This difference is best described in the three theses from Heidegger's famous seminar *The Fundamental Concepts of Metaphysics: World, Finitude and Solitude*: "The stone is worldless, the animal is poor in world, the man is world-forming."[7] Having or not having language is thus visible in the way the three different types of beings relate to the world, three different ways the world is given to them, three different ways in which they have access to beings. The latter expression is used by Heidegger only as a provisional means of explaining what the world is,[8] but is nevertheless a close approximation for the point he is making.

If we take Heidegger's preliminary definition of world, we can define the worldlessness of the stone as not having access to beings, a not-having that does not entail any having-had.[9] The examples given by Heidegger make what he means quite clear: the stone lies on the ground, there is contact between stone and ground, but we cannot say that the stone is *touching* or *feeling* the ground, for this would entail that the stone adopts a certain stance towards the ground. It is obvious that the stone does not see, feel, smell, hear or taste anything in the world, nor does it have preferences – a stone thrown into a lake or to the side of the road does not like or dislike its new surroundings (much as it did not like or dislike the old ones).

An animal clearly is a different case altogether. Heidegger explains the difference between the stone and the animal with the example of a lizard.[10] The lizard feels the stone, even if not *as* a stone; it feels the sun; it recognizes being removed from a place it chose and strives to return to it. It definitely has senses and preferences, and tries to change its position in respect to what we would classically call "world." This is clearly different than in the case of the stone. From the behavior of the animal we might guess that it has a certain type of access to beings, and that, contrary to what was said earlier, seems to indicate that animals do in fact *have* world. Indeed, Heidegger notes this fact in observing that the analysis summarized previously suggests that the animal "both has and does not have world."[11] It certainly has some (if limited) access to beings, so in comparison to the stone, it does have a type of "world." But, in comparison to man, it does not have world, because, as we will see in some time, a man's world is different from the one animals (do not) have. The animal type of access to beings is thus different from that of man, and if man's type of access to beings is what is called "having world," then the animal has no world.

To describe the difference between the human and the animal access to beings, Heidegger introduces another pair of concepts. The animal type of access can be described as *behavior*, while the human one, as *comportment*. Animal behavior is constituted by repetitive, instinctual reactions that are made possible by captivation by its environment. This is followed by two well-known examples regarding bees. In the first one, we see that even though a bee seems to "know" how much honey or pollen[12] it needs to eat or gather from a flower before it can go back to the hive, this "knowledge" is in fact based on a simple reaction to a fullness of stomach – a bee with a severed abdomen will continue sucking, not noticing that anything is wrong.[13] This points to the fact that a bee does not understand the honey *as* something that fills (and can potentially overfill) its stomach.

The second example is based on a seemingly more complicated behavior: namely, the way bees find their hives even after having covered a large distance. Introducing certain changes in the bees' environment reveals much about the way they do it: if a hive is moved just a few meters, the bees will first fly blindly to the spot where it used to be (even though one might expect they see it perfectly, as they are capable to discern their hive from others by its color), and only then wander around to try to find it. If a bee is caught outside its hive and released later in the day, it will go astray. If a caught bee is released in the opposite direction of the hive than the one it was caught on, it will fly in the opposite direction than it should. All this can be explained by the fact that bees have a sense of the distance they covered and use the sun for direction. But what do all those behaviors tell somebody that searches not for reasons for, but the essence of behavior?

Heidegger seems to argue that a bee makes "mistakes" that no human being would make – or at least in a way no human being would make them. Let's take the example of the sun. It does not take a scientist, or even a modern, post-Copernican man to understand that the sun changes its place with respect to the horizon, that it is "moving." A man would not blindly follow the exact same angle the sun was forming when he left; when he woke up, he would not start walking east instead of west, because he was walking "into the sun" when it set. The point I am making is not one of astronomy – a man can know nothing about what the sun *is* and still not behave in this way. This is because a man apprehends the sun *as* sun, that is, for example, a flaming ball of light that moves around the sky, and not just as another source of light or a static signpost that somehow disappears sometimes. If a man would be mistaken in finding direction according to the sun it would be because his understanding of astronomy is wrong, but he would still comprehend the sun *as* sun.

For Heidegger, the animal is certainly open towards the world, as it receives stimuli from outside, but to say that it is open purely and simply would be wrong. In fact, the animal is "absorbed within itself," it "remains within itself."[14] This is what Heidegger calls "captivation." The animal, as enclosed in itself, cannot go out of itself and see itself as if from the outside. That is why the bee cannot see that it has eaten too much honey – it cannot see itself as a vessel with a finite capacity for eating. It also cannot see its current situation from the outside, which leads to all the mistakes in direction described before. Most importantly, the bee cannot see beings as a whole from the outside; that is why it does not have world.

Man has language, he is a *zoon logon*; that means that he *is* in a certain way. We have come to see that this way of being can be described as "having world" or "being world-forming." The animal does not have world in the mode of being poor-in-world; this stems from the fact that the animal is absorbed in itself. To have world, man must thus not be absorbed in himself in the way described previously.

This essential trait of man is based on an ability to "leave" himself, to "project" himself, as Heidegger puts it in *The Fundamental Concepts*. This projection is temporal – only understanding himself as a temporal being can man go "outside of himself" and see things in their "as such," thus "having world." In *On The Essence of Ground* this is referred to as *transcendence*.

In our comparison between man and animal, we have seen how man transcends his situation, which lets him see things *as* things. But these were only things that were "present at hand"; to understand Dasein in its essence, we must go a step further, and say that Dasein as Dasein transcends the world as such as a whole.

It is important to note that for Heidegger, transcendence is not a remote being that Dasein would have access to by transcending the world and going outside it; it is certainly not God. Transcendence is the original movement that allows Dasein to be Dasein and for the world to be world, understood as "manifestness of beings," or the access to beings as such. Heidegger:

> "Dasein transcends" means: in the essence of its being it is *world-forming*, "forming" [*bildend*] in the multiple sense that it lets the world occur, and through the world it gives itself an original view (form [*Bild*]) that is not explicitly grasped, yet functions precisely as a paradigmatic form [*Vor-bild*] for all manifest beings, among which itself Dasein belongs.[15]

The transcendence of Dasein thus allows the formation of world as a paradigm, as a transcendental possibility of things to be apprehended *as* things – and to see itself as one of them. Transcendence, as I said, is a temporal means of Dasein seeing itself in its essence (which is the essence of man, strictly speaking). But to see this essence, Dasein must understand itself as finite. Thus, the transcendence of Dasein can only be accomplished by establishing a rapport to its own death.[16]

In a typical move, Heidegger rejects any notion of death that has anything to do with the notion of life – certainly, everything that lives dies, but life is a phenomenon of biology (thus an ontic and not ontological one) and the analysis of Dasein will result in a notion of death that is primary to it.[17] He also opposes to any theory that would operate in the "this world-that world" logic, as Dasein is precisely in the mode of having-(this)-world.[18]

Heidegger is not talking about the experience of death – neither our own or someone else's.[19] Death is certainly the end of Dasein, and thus something that allows it to see itself in its totality and finitude. However, this end is not a debt to be repaid in days or whatever other currency, nor is it a growing of Dasein to a prede-termined state, which is only accidentally unreached (it is not like the moon which "grows" only to reach its actual size). Nor is it an act of ripening – Death is not the goal of Dasein, even though it goes out towards it in a way, and even if it makes it a totality. None of the usual conceptions of the end fit this notion of death – it does not end as the rain that stops dropping, it does not end like (paradoxically) an unfinished road, nor does it end like a painting, with the last brush stroke: all these conceptions of the end can only be ascribed to things ready-to-hand.[20]

Death is rather a way of being, and Dasein's essential way of being is being-towards-death.[21] The most complete, positive definition of this way of being is such:

> death, as the end of Dasein, is Dasein's ownmost possibility – non-relational, certain and as such indefinite, not to be outstripped. Death is, as Dasein's end, in the Being of this entity towards its end.[22]

Let me comment shortly on these two dense sentences. Heidegger has found in death what he had been looking for – something which allows Dasein to be itself individually ("ownmost possibility") as a whole (totalizing it as its end). Death is non-relational, which means that it effaces any relations to other Daseins. Saying that it is "certain and as such indefinite" Heidegger not so much pronounces a banality (everybody dies, but nobody knows when) than discovers the presup-position of this banality (everybody dies <but me>, but nobody knows when <but I make plans for how I am going to die>) and remedies it, again by individualiza-tion: *I* as Dasein am going to die, and I do not know when (this is one of the causes of *anxiety*). It is not to be outstripped, because there is no possibility of Dasein that could surpass it, or that could come after it – in fact, there will be no possibil-ity of Dasein after death, which is the (only) possibility of Dasein's impossibility.

Death, in Heidegger's existential-ontological view, is only insofar as it is a death of an individual Dasein. Dasein, as I said earlier, exists only in the tran-scending of the world as such as a whole, thus seeing it in its totality – and itself in its totality in death. This means that Dasein needs death as a point to which it tran-scends in order to essentially be Dasein. If Dasein, in its essence, is the essence of man, it means that man, and only man, is capable of this transcendence towards death, which, in fact, is death itself (death, in Heidegger, is being-towards-death). Indeed, the animal – according to Heidegger – does not die.[23] The animal does "perish," "come to an end," "cease to live" – however, it does not die in the Hei-deggerian sense.

To summarize: The definition of man as *zoon logon ekhon* describes man's specific way to be, which is characterized by language. Language, in turn, is a way of transcending the world as such as a whole. This transcendence is possible only when Dasein can see itself in its own finitude, which is only possible *via* a special relationship to death. As Joachim L. Oberst scrupulously shows, three of Heidegger's seemingly unrelated theses – that only humans *exist* in the proper sense of the term (i.e., transcend) that only humans "properly" die, and that language is properly human – are indeed tightly interconnected.[24] And, as the analyses conducted previously have shown, all three can serve as the location for the human–animal caesura – and so does thinking, alluded to in the beginning of this section, which is, for Heidegger, secondary to language, and is closely linked with the hand and handiwork. This link will be crucial in the last chapter of this book.

What is important to note is that for Heidegger man is truly human – is Dasein – only insofar as he transcends. This has two key implications for the task undertaken in this book. First, it shows that Heidegger's concept of man is created on a plane of transcendence – while Heidegger does not talk about God or anything out-of-this-world, it is clear that in order to actually be Dasein, in order to have language or think, man has to transcend the world, using death as an external point of reference – and at this point a plane of transcendence is established.

Secondly, it is important to see how in Heidegger man is not always human – he is so only insofar as he transcends. As Didier Franck comments, "The extatic determination of man's essence [i.e., that man is human only insofar as he transcends – K.S.] implies the total exclusion of his live animality, and never in the history of metaphysics has the Being of man been so profoundly disincarnated."[25] This does not mean that Heidegger sees man as a purely disincarnated being – on the contrary, man does have a certain "live animality" about him, but it is not his essence. Man is thus both human (Dasein) and animal ("live animality"), while the animal is never transcendent. Moreover – and this will be crucial in the next chapter – the animality of the animal can only be understood in reference to the humanity of man. And although between *Sein und Zeit* and *The Four Fundamental Concepts* . . . Heidegger seems to have moved from a simply privative understanding of animal life with respect to Dasein to a more complicated view.[26] The very notion of poverty-in-world and the fact that it is explained in terms of the inability to comport towards things *as* things is already very telling. In fact, one can understand the complicated mechanics of the disinhibition ring as the answer to the question: "If animals are not capable of having world, how can they act within it in such a complicated fashion" – analogous to Descartes' question: How can they behave in such a complicated manner if they do not have souls?

This theme – man as a transcendent being (or one capable of transcendence) which also has an animal side, and animal as an immanent being incapable of transcendence – will repeat itself in the next examples.

A note on Agamben

Despite the originality of his interpretation, Heidegger upholds a very classical definition of man as a *zoon logon ekhon*, the animal that is not animal because it has language – which is one of the reasons his work was chosen for this study – but it might be interesting at least to mention a theory which, although it upholds a radical difference between man and animal based on the possession or non-possession of language, is, at least in appearance, highly non-classical in this respect. The theory is that of Giorgio Agamben, put forward in *Infancy and History*.

Agamben sees human uniqueness in the fact that only men are ever infants – which means that they are, in an ontological as well as chronological sense, temporarily deprived of language in its true, semiotic form.[27] The semiotic aspect of language is that of signs and their correspondence to objects. In order to function in the semiotic aspect of language, one needs to recognize this correspondence – when hearing the word "horse," recognize that it pertains to a real horse. The semantic aspect is that of meaning, of producing messages that cannot be reduced to a mere succession of semiotic signs, and are not produced as a simple addition of signs. Instead, it is the meaning itself, the universe of discourse "in its total conception, which is enacted and which divides itself into specific 'signs,' which are words."[28] Functioning in the semantic aspect of language, grasping meaning, requires understanding, which is more than just recognition of signs and correspondences and requires the ability to refer to the higher order of meaning.

For Agamben, the animal type of existence in language is purely semiotic. Animals are always already in language; they do not recognize themselves as distinct subjects, because they know nothing of being outside – Matthew Calarco, in commenting this fragment, uses Bataille's expression "water in water."[29] A man, in every act of speaking (which Agamben sees as going into the semantic order), which entails constituting himself as a subject by saying "I," momentarily removes himself from the semiotic, from the immersion in language: "Like dolphins, for a mere instant human language lifts its head from the semiotic sea of nature."[30]

This theory, in fact, is exactly opposite to the traditional view held by many philosophers from Aristotle to Heidegger and onward:

> Contrary to ancient traditional beliefs, from this point of view man is not the "animal possessing language," but instead the animal deprived of language and obliged, therefore, to receive it from the outside himself.[31]

Indeed, if a child is not exposed to language early enough, it will never be able to speak properly. However, as we know thanks to Chomsky and Lenneberg (whom Agamben cites),[32] human language is not only acquired from the outside, but also partly innate. It is this double feature of human language (semantic and semiotic, esosomatic and endosomatic), which leads Agamben to say that man is on the edge between the semantic and semiotic, esosomatic and endosomatic:

> It is this position *on the boundary* between two simultaneously continuous and discontinuous dimensions which makes human language able to transcend the purely semiotic sphere and to acquire, in Benveniste's terms, a "double signification."[33]

However, a closer look at Agamben's theory will reveal that it is not as distinct from the traditional approach as we might think at first glance. What infancy – the fact of not-always-being-in-language – allows for is the transcendence of the animal state of being "water in water" – "lifting our heads from the semiotic sea of nature." Exactly as in the case of Heidegger, what makes men different from animals is this ability, the movement of transcendence. And here, too, it requires an ability to refer to an external point – it is the totality of "meaning," identified by Benveniste.

And, equally importantly, man is not just human (not just operating on the semantic level), but also animal – he is, as we said, on the edge, or perhaps *is* the edge itself. But what is truly human is the precise ability of transcendence. Once again, the animal remains locked inside the immanence of "water in water" while the human transcends it thanks to an ability of making an external, absolute reference.

Freud and the political animal

The choice to analyze the work of Freud at this point may be deemed problematic – the father of psychoanalysis does not say a lot about animals, nor does he explicitly use the definition of man as *zoon politikon*. However, there are at least two reasons to include him in this study.

First of all, Freud is a post-Darwinian thinker who took evolution seriously – in fact, in a frequently quoted fragment, he likens psychoanalysis to Darwinism explicitly in his famous evocation of how both sciences dealt blows to the "*naïve* self-love of men."[34] The fact that Freud knew Darwin's theory and treated it seriously should make us especially attentive to the way he understands the difference between men and animals and the logic of its functioning, and so should the fact that Freud was a declared atheist, which makes any simple theological interpretation impossible.

Secondly, even though he does not use it explicitly, Freud in fact upholds the classical definition of man as a *zoon politikon*. My analyses on the next few pages will show it more clearly; for now, let me treat this fragment of the introduction to *The Future of an Illusion* as preliminary proof:

> Human civilization, by which I mean all those respects in which human life has raised itself above its animal status and differs from the life of beasts – and I scorn to distinguish between culture and civilization – presents, as we know, two aspects to the observer. It includes on the one hand all the knowledge and capacity that men have acquired in order to control the forces of nature and extract its wealth for the satisfaction of human needs, and, on the other hand, all the regulations necessary in order to adjust the relations of men to one another and especially the distribution of the available wealth.[35]

I would like to stress the second part of the definition: if what distinguishes men from animals (I will show in a moment that this difference is both essential and qualitative) is culture or civilization, then men have a specific system of organizing relations between themselves which is neither present in nor available for animals.

Why can this relation be called political, and not simply economical? Because men are not purely rational creatures, and only this type of creature would be capable of such a simple trade relation. The difference between a "state of nature" (which may or may not have ever existed in the first place) and "civilization" for Freud is that in the former every individual would act in order to satisfy their basic instincts as quickly and as fully as possible, and would not resign from any wealth he or she could acquire unless forced to do so by another such individual. Culture forces individuals to resign from satisfying some of their instincts for the good of the community – one of these resignations is certainly not acquiring too great a deal of wealth, be it measured in money, sex or any other currency.

However, this is only a description of culture, and not its essence. The latter can be most clearly shown by exposing the origins of culture. In this work I am not interested so much in Freud's possible contributions to psychology or sociology, as, again, the logic of his thinking or its premises. The question is not "how was it possible that culture started developing," but "what is culture, what is the human *polis* (in opposition to an animal herd) in its beginning, in its essence."

The beginning or essence of culture or of the *polis* is, interestingly enough, a doubly mythical one – it recurs in many myths and tales, from the different variations of the story of the death of (a) god (not only the crucifixion of Jesus, but also Egyptian myths), to the murder of the father in *The Brothers Karamazov*; and more importantly, it can only be told as a hypothesis, a primary state that could have or even must have logically occurred, but cannot be proven neither by ethnographic analysis of the most "primitive" societies we can now observe nor by archaeological or historical data. The only observation Freud possesses is the existence of totemic bands of brothers.[36]

The genealogical explanation Freud provides for the shift between the supposed primal horde and the existing totemic societies is well known. The primal horde is ruled by a jealous father, a strong male who forbids his sons to have sexual intercourse with any of the women. One day, the enraged sons kill and eat the father in order to end his rule. However, after the deed they were horrified – as, apart from hate and frustration, they also felt love for their progenitor – and repented by putting forth two primary commandments: the prohibition of patricide, which then became the prohibition of the killing of the totemic animal outside of rituals performed once a year in cathartic commemoration of the initial deed, and the prohibition of marriage inside their group, which would ensure that no such jealousy as that which produced the initial murder could ever take place. In the words of Freud:

> For a long time afterwards, the social fraternal feelings, which were the basis of the whole transformation, continued to exercise a profound influence on the development of society. [. . .] It was not until long afterwards that the prohibition ceased to be limited to members of the clan and assumed the simple

form: 'Thou shalt do no murder.' The patriarchal horde was replaced in the first instance by the fraternal clan, whose existence was assured by the blood tie. Society was now based on complicity in the common crime; religion was based on the sense of guilt and the remorse attaching to it; while morality was based partly on the exigencies of this society and partly on the penance demanded by the sense of guilt.[37]

What is important in this fragment is that although it was indeed "fraternal feelings" that made the transformation from the primary horde to the band of brothers possible – as without a feeling of communality of their fate they could not have acted together to kill the father in the first place – those feelings were not enough to ensure the beginning of civilization as we know it. What is needed is the double movement of internalization of the father's laws – especially "Thou shalt not kill" (first the father, then the brothers, then anyone at all) – and ensuring the existence or the memory of an external source of these laws: the father and the terrible crime committed upon him. Only the existence of this reference and the ability to form it can ensure the creation of actual culture or the *polis*, and, I can add, it is exactly this reference that differs men as *zooi politikoi* from animals.

What plays out in the story of the murder of the father out of desire to women (who were collectively mothers of the tribe), is what, on the level of the individual, is of course known as the Oedipus complex. It is not the place here to give a detailed account of the Oedipus complex.[38] I will proceed straight to the most important interpretations. The phase of the Oedipus complex and its fall is crucial for the entering of the child into the "civilized" world, where a boy learns to sacrifice some of his desires (the most important of which are the desire to have intercourse with the mother and to remove the father who is an obstacle to the fulfillment of the child's incestuous plans) for the good of social being. It is interesting to note in this context that children in an early phase consider themselves more alike to animals than to grown-ups[39] – we might say that it suggests that growing up and becoming part of the *polis* is in fact moving from the animal to the human world. In order to move to that world, children must internalize the initially paternal prohibitions. However, this internalization introduces a splitting of the child's ego, a part of which now becomes the partly unconscious superego. The ability to form a reference to the superego is essential for any civilized behavior, i.e., for being part of the *polis*. This way, the process of civilizing or indeed "humanizing" the young half-animal includes forming a plane of transcendence right in the middle of his or her *psyche*.

Similarly, the more practical side of civilizing, i.e., forming groups, also requires an ability to form a relationship with an external point of reference. As Freud says in *Group Psychology and the Analysis of the Ego*, in reference to groups such as the army and the Church:

> A primary group of this kind is a number of individuals who have put one and the same object in the place of their ego ideal and have consequently identified themselves with one another in their ego.[40]

Thanks to the reference to an external being – either empirically present as in the case of the general in the army, or present only "spiritually" as in the case of God or Jesus in the Church[41] – the individual can form a relationship to other individuals and see them as "brothers," as members of the same community.

Let me summarize. In totemism we saw the historical essence of the *polis*, which was the forming of a bond with an external point of reference to the dead father or the deed of murdering and eating him. In the fall of the Oedipus complex, we saw the ontogenetic basis of entering the *polis*, which was the introduction of a transcendent referent into the ego. The short analysis of groups revealed this reference as the key to forming human societies of different kinds. All three lead to the same conclusion – what is human in man, understood as *zoon politikon*, is functioning on a plane of transcendence, or the ability to form external references to absolute laws or figures.

A note on Schmitt

Carl Schmitt seems an even more problematic figure for the analysis of the *zoon politikon* definition than Freud – his work has little to do with any aspect of animal studies, and aside from Jacques Derrida's seminars on *The Beast and the Sovereign*, he is rarely mentioned in the context of animal philosophy. There are, however, several reasons why an engagement with Schmitt's thought is important for the understanding of the difference between man and animal in the realm of the political.

Firstly, Schmitt's political thought has been an important point of reference – positive or negative – in 20th-century political philosophy for figures such as Slavoj Žižek, Chantal Mouffe, Ernesto Laclau, Michael Hart and Antonio Negri, or the aforementioned Jacques Derrida. This shows that Schmitt a figure to be reckoned with and his thought may, to a certain extent, serve as paradigmatic.

Secondly, Schmitt sees the political as an absolutely autonomous domain, inherently different from other "endeavors of human thought and action, particularly the moral, aesthetic, and economic."[42] This means that the definition of *zoon politikon*, if used – as I will try to show, implicitly – by Schmitt, should not rest upon another domain, for example that of biology or metaphysics (although, as we will see, it is linked closely to a certain anthropology). This means that if indeed man is the only animal capable of forming political communities, he is so because the political rejects the animal as such, and not because of a contingent reason resulting from the misuse of antiquated biology or metaphysics (for example a misunderstanding of Darwin's theory).

Thirdly, as I will try to show in some more detail, Schmitt's thought is a perfect example of an almost explicitly theological plane of transcendence. As the German jurist famously asserts in *Political Theology*, "All significant concepts of the modem theory of the state are secularized theological concepts not only because of their historical development [. . .] but also because of their systematic structure."[43] This is especially important for Schmitt's decisionist thought, wherein he distinguishes the sovereign as "he who decides on the exception"[44]

– the jurisprudential exception is likened to a miracle in theology, entering the natural order of laws from the outside, and yet functioning as their condition. The miraculous exception of the sovereign decision is the transcendent referent which governs the political functioning of a community.

The theological character of politics does not, in itself, prove that it is, for Schmitt, a purely and essentially human matter. The proof is needed all the more in the light of *The Beast and the Sovereign*, where Derrida posits an equivalence – or at least analogy – between the animal or beast and the sovereign or the law-giver: "The law is always determined from the place of some wolf."[45]

However, while both the beast and the sovereign remain outside of the political realm, it seems that their "outsideness" is of a different nature – forming a limit to Derrida's analogy. While the sovereign, by virtue of his decision, extricates himself from the political realm in order to ensure its existence as a transcendent referent, the animal is incapable of entering the political realm in the first place, rejected by the very formula of the political.

While the truly sovereign decision, for Schmitt, is the declaration of the state of exception, a fundamentally political decision concerns the difference between the friend and the enemy. The declaration of enmity is, at the same time, the forming of the political community, which can only exist as political insofar as there is an enemy, an outside of the community. In this sense, the political encompasses the formulation of a plane of transcendence, wherein a given community or state creates its political identity only in reference to another community or state (the enemy), at the same time rejecting everything that is not capable of creating such an identity for itself or for the other (i.e., becoming a friend or enemy) outside the realm of the political.

This rejection is especially visible in the analysis of wars waged in the name of humanity. As Schmitt notes: "there are no wars of humanity as such. Humanity is not a political concept, and no political entity or society and no status corresponds to it."[46] It is so because "[h]umanity as such [. . .] has no enemy, at least not on this planet."[47] Humanity, at least in the current state of our knowledge of the Universe, is in itself outside politics, because any political friend–enemy combination must come from within it. The political gesture is based on the work of the anthropological machine in the Agambenian sense of the term – before naming an enemy, the community always already knows that an enemy must come from within humanity.

What is more, a war waged in the name of humanity – or, rather, conflict in the name of humanity, as it would not, strictly speaking, be a war – ceases to be political and dehumanizes the opponent: "Such a war is necessarily unusually intense and inhuman because, by transcending the limits of the political framework, it simultaneously degrades the enemy into moral and other categories and is forced to make of him a monster that must not only be defeated but also utterly destroyed."[48] When the enemy is made into an enemy of humanity, he becomes something "less" than human: the monster, the beast, the animal.

This necessary humanity – in the classical sense of the term – of the enemy is also important for Derrida's reading of Schmitt in *Politics of Friendship*. While analyzing Schmitt's refusal to use the term "humanity," Derrida comments on the jurist's "phallogocentrism" and the rejection of women from the political.[49] This points to Schmitt representing a classically humanist paradigm, in which the supposedly universal, "human" concept of the *zoon politikon* rejects anything that does not conform to a much more restricted view of "humanity." And, as we have seen in Chapters One and Two, this kind of rejection is one of the staples of the creation of the concept of the animal. I will come back to this in "False Immanence."

To sum up: Schmitt's theory shows that the political needs a plane of transcendence to be established. This can be done either (1) on the level of the sovereign decision, which, analogically to the miracle in theology, always comes from the outside of the order of established laws; or (2) on the level of constructing the identity of the political community, by means of the declaration of enmity. Especially in this last instance, the establishing of the plane of transcendence is only possible within humanity, thus rejecting the animal from the political. While the animal cannot achieve the level of the political (it can be neither friend nor enemy), men can be rejected from the political, as evidenced by the case of the supposed "war" in the name of humanity, in which the enemy is dehumanized – effectively turned into "a monster." This way, while the animal stays the apolitical *zoon*, the man is human only insofar as he is within politics, outside of which he becomes an animal.

Kant and the *Animal Rationale*

Including Kant in this chapter seems to break with the rule of only speaking about post-Darwinian thinkers, which ensures that the understanding of the human–animal distinction is not biased with a lack of biological knowledge. However, the Kantian approach to ethics remains an important point of reference – positive (if reservedly) for the proponents of different variants of "rights" approaches, and negative for utilitarians. Indeed, when animal studies theorists analyze Kant, it is usually to focus on the immediate consequences of his ethics for animals and/or to show how the Kantian approach can or needs to be amended in order to account for our duties towards animals. The engagements with the German philosopher range from wholesale rejections to attempts of showing that while Kant explicitly rejects that we have any direct duties towards animals[50] (driven by the means–ends distinction already present in Aquinas), the spirit of his thought suggests more favorable options.[51]

While not underestimating the importance of the purely ethical aspect of Kant's thinking, I would like to focus on the logic of the difference between man and animal as present in Kant and the planes of thought that are in play here. As I will try to show, what is decisive for the animal–human caesura in Kant's philosophy are certain pivotal fragments of his thought that do not deal explicitly with animals, but inform the way in which the division operates.

For Kant, man is a rational animal. What is important to note in the beginning is that the German philosopher does not determine if man is the *only* rational being in the universe – in fact, the *Anthropology* includes quite a few interesting discussions of extra-terrestrial intelligent beings.[52] And indeed, Kant's theses from the *Critiques* are as valid for man as they are for any other rational being – it is hard to say that he is talking about reason or understanding in terms of an "essence" of humanity that would make it radically different from all the other beings. However, it is particularly telling that Kant considers the difference between man and animal so vast that he would search the skies above him for a rational being rather than look at an ape or a crow.

In the *Anthropology*, Kant proposes the following basis for the human uniqueness in the animal realm:

> The fact that the human being can have the "I" in his representations raises him infinitely above all other living beings on earth. Because of this he is a person, and by virtue of the unity of consciousness through all changes that happen to him, one and the same person – i.e., through rank and dignity an entirely different being from things, such as irrational animals, with which one can do as one likes. This holds even when he cannot yet say "I," because he still has it in thoughts, just as all languages must think it when they speak in the first person, even if they do not have a special word to express this concept of "I." For this faculty (namely to think) is understanding.[53]

As in the case of Heidegger, for Kant thinking is close to language, but here their relationship is situated differently – saying "I" is first and foremost a question of thinking, and only secondly of actually pronouncing the word. The logic of language follows the logic of reason.

The last sentence of the cited fragment, invoking understanding, indicates that Kant is talking about speculative reason. More importantly, the whole passage alludes to what the *Critique of Pure Reason* describes as the (transcendental) unity of apperception, the transcendental ground for the fact that all *our* experience is viewed exactly as that – experience belonging to one consciousness, one person, even though at different points in time.[54] Different experiences are united as happening to one person. How exactly is this unity linked with understanding?

The transcendental unity of apperception makes cognition possible by subjecting all experience to categories (concepts of pure understanding, i.e., *a priori* conditions of links between phenomena).[55] This means that it is thanks to the transcendental unity of apperception that we experience phenomena as connected in a regular or necessary manner.[56] The phenomena are linked not because we perceive real connections between noumena – things in themselves – but because our perceptive apparatus only allows for such and such connections.

This receptive – even though active in its receptivity – power of apperception is especially important for empirical cognition. Thanks to it, the human being can observe nature as a causal system in the sense of empirical (and not final)[57] causality, and consider the cause-effect links as necessary, thus observing nature scientifically.

When we start doing so, we notice that there is a chain of causes and effects that leads far into the past, and start wondering whether there is a supreme, primary cause, which would be both the starting point and the guarantee of the necessity of nature's order. In a word, we start wondering if there is a God.

Kant famously rejected all the so-called proofs of, or arguments for, the existence of God – the ontological (God's existence is proven by the analysis of the concept of God), the cosmological (the existence of the world implies the existence of its creator) and the physico-theological (the world is showing signs of a purposive design, therefore there must be a designer). While the details of each argument remain irrelevant to my reasoning, I would like to note that the underlying premise of each rejection is the inability for pure reason to exit the confines of experience, that is, to cross the phenomenon–noumenon barrier and reach out into transcendence, where such a God would reside. However, the existence or non-existence of God ultimately turns out to be irrelevant to Kant.

What *is* relevant is that the metaphysical idea of God (along with the idea of an immortal soul and the world being a causal whole) is both natural and necessary for our empirical endeavors. Kant calls all those three (God, immortal soul, world-whole) "regulative principles" – not necessarily existing, but guiding the use of reason. One might say that they are what governs the scientific "plane of thought," according to Kant. The usefulness of the idea of God, in particular, is

> to regard all combination in the world as if it arose from an all-sufficient necessary cause, so as to ground on that cause the rule of a unity that is systematic and necessary according to universal laws; but it is not an assertion of an existence that is necessary in itself.[58]

Again: the existence of the highest being is irrelevant; however, as humans, in order to think (and speak) speculatively, i.e., to use our faculty of understanding, we must think *as if* there were a supreme being.

The same is true when it comes to practical reason. Kant's argument can be summarized thusly: When man turns his speculative gaze to himself, at first he notices that, like the rest of nature, he is a causally driven object. He is subjected to the same laws and necessities as the rest of nature; his body is an object that undergoes the same processes as animals. Of course, this speculative gaze, allowed by understanding, in itself makes man different from animals. Yet, there is another power – the one which allows him to *turn* the gaze wherever – that is also closely linked to the unity of apperception – reason.

In the "phenomenological" reasoning presented previously, the discovery of reason is made via pure apperception – the human being, thanks to his ability to analyze his perception as belonging to himself, notices that he is *not only* one of the causally driven objects, but also is capable of other behaviors, for example the aforementioned directing the understanding to such or such objects. The faculty which allows him to do so is practical reason, which enjoys primacy over speculative reason.[59]

Practical reason has the directive force when directing what I have called the speculative gaze of understanding, because it is driven by practical interest.

However, reason's practical interest does not lie primarily in speculative matters, but in moral ones. While it might be obvious in the speculative, purely cognitive use of reason, in its moral use it is crucial to underline the fact that this determination is not based on any appetitive, "animal" – as Kant would say: "pathological" – feelings, but is absolutely free from them. This moral determination thus entails a certain concept of freedom.

Practical freedom is not the simple ability to choose – animals choose too, having "*arbitrium brutum*"[60] – but it is the ability to choose regardless of the constraints of sensibility. For example, an animal as well as a man can choose to defend its children up to the point of sacrifice. However, this is not a free choice – it is "pathological," driven by the love for the children or the sheer instinct to protect them. It is the "sensible impulse" that drives the *arbitrium brutum*. A really free choice, an act of free will – *arbitrium liberum* – is subjecting our choice to practical reason and it alone. Kant:

> Freedom in the practical sense is the independence of the power of choice from necessitation by impulses of sensibility. For a power of choice is sensible insofar as it is pathologically affected (through moving-causes of sensibility); it is called an animal power of choice (*arbitrium brutum*) if it can be pathologically necessitated. The human power of choice is indeed an *arbitrium sensitivum*, and yet not *brutum*! but *liberum*, because sensibility does not render its action necessary, but in the human being there is a faculty of determining oneself from oneself, independently of necessitation by sensible impulses.[61]

We have already seen that one of practical reason's ways of acting is determining the areas of interest for speculative reason. However, the area where practical freedom as liberated from sensibility truly shows itself is course that of morality. How does it work? In phenomenological terms, close to what I have indicated earlier, one might argue that a man finds among his actions certain that seem not to belong to the realm of natural necessity. These actions, rather than being effects of internal or external causes, are directed towards an end. Among them, a special category is that of moral actions – directed to moral ends. What really makes up a "moral action" is described in Kant's *Groundwork for the Metaphysics of Morals* as an action driven by the categorical imperative. There are three forms of the categorical imperative: the first one focusing on the universality of the action, the second one making us treat our action as if it were to become a law of nature, and the third one saying that it should use the humanity in ourselves and others not only as a means, but also as an end.[62] Now, every one of these formulations focuses on a different side on what is moral. The first one shows that a moral action would be impossible without the unity of apperception – not only can man say I, but also, he is capable of seeing that others are similar individuals, capable of saying I and having different wants and needs. The second formulation of the categorical imperative teaches us that also in moral laws, teleological rather than natural, there is or should be a necessity. The third, finally, focuses on the fact that what one is looking for is the final end of an action, and it is always humanity.

If there are necessary ends (as there were necessary causes), we need to assume that there is a hierarchy among them, and that one is the most important, the most necessary. As Kant says:

> Consequently, we must assume a moral cause of the world (an author of the world) in order to set before ourselves a final end, in accordance with the moral law; and insofar as that final end is necessary, to that extent (i.e., in the same degree and for the same reason) is it also necessary to assume the former, namely, that there is a God.[63]

Here also, as in the former example of natural necessity, the final conclusion, namely that there *is* a God, says little about the actual existence of that being, but rather shows it as a regulative principle of our reason. This, again, means that we do not need to know that God exists, but in order to think morally, we need to think *as if* he existed. As Kant continues in the footnote:

> This moral argument is not meant to provide any objectively valid proof of the existence of God, nor meant to prove to the doubter that there is a God; rather, it is meant to prove that if his moral thinking is to be consistent, he must include the assumption of this proposition among the maxims of his practical reason. – Thus it is also not meant to say that it is necessary to assume the happiness of all rational beings in the world in accordance with their morality for morals, but rather that it is necessary through their morality. Hence it is a subjective argument, sufficient for moral beings.[64]

In moral thinking, God is thus a regulative principle, which serves as a necessary anchor in our thinking: for moral thinking to be consistent, in fact for it to exist at all, we need to assume the existence of God.

Let me summarize where we got thus far. Kant's remark from the *Anthropology* suggested that it is the "I" that we have in every thought and representation that distinguished men from animals, which do not possess this trait – in its pure form called the transcendental unity of apperception. This trait allows us, on the level of speculative reason, to see the link between different phenomena as necessary – that is to subject them to categories, for example of causation. This leads us to believe that there is a necessary first cause, which is called God. While proving the existence of this first, necessary being lies outside the possibilities of our reason, nevertheless we have to think *as if* it existed for our (scientific) thinking to be possible.

When we turn the speculative gaze to ourselves, we can also analyze ourselves as one of nature's beings, subjected to the same laws as other such beings. However, we also notice that there are at least two actions our reason is capable of that escape this simple causation – the very act of turning the attention of speculative reason to this or that problem, and moral actions. Both of them are free in the practical sense (proving the existence of theoretical or "natural" freedom is beyond reason), which means that they are *a fortiori* not "pathological," that is they are not guided by any kind of sensual or empirical interest, but by reason alone. This guiding reason is dubbed "practical" by Kant.

However, when analyzing moral choices, the most important of those which we make using practical reason, we see that they too are driven by necessary laws. These laws are not causal, like it was with natural ones, but they are driven by an end. As in the case of natural causation, to think those moral laws consistently, one must think that there exists a supreme, final end. Or, more correctly, one must think *as if* such a final end existed, because – again – the proof of its existence is impossible, as it lies beyond the reach of speculative reason.

Man differs from animals in the sense that it can think nature as a system of necessary causes, and that it can think morality as a system of necessary ends. For both of these to be possible, man, as human, needs to think *as if* there were a prime, necessary cause and a final end.

When it comes to the positive appraisal of the animal, Kant does not seem to have much to offer. Animals are simply animate parts of the natural order. Not possessing understanding, they are unable to view themselves as parts of nature; not possessing reason, they are unable to act morally and thus are not worthy of our respect – as said earlier, only rational beings deserve respect, according to Kant. They are thus always in the grips of their sensibility – never free.

Most importantly, the fact that animals are not endowed with reason (or understanding) means that they are unable to form a thought relationship with a transcendent God. Man is only human insofar as he forms this kind of relationship, even though the object of this relationship needs not necessarily to exist.

Reason or understanding might be considered the distinguishing trait of humanity by Kant, but only after this analysis one can truly understand what that means. More than in terms of essence, to understand Kant's way of explaining the difference between man and animal, one must think in terms of ways of being (or indeed "ways of thinking"). Man is a being which is (or thinks as) human only thanks to a relationship to the necessary cause of nature and the final end of morals. He needs the ability to direct his thought to the unknown transcendence in order to actually be human. The animal is deprived of this possibility.

In the chapter on planes of immanence and transcendence, we have already seen that Deleuze and Guattari count Kant among those thinking in terms of the latter plane. However, his short comment is not immediately related to the animal–human caesura, as according to *What is Philosophy?* the fault of Kantianism, or transcendentalism in general, is that it replaces the transcendence of God with "that of a Subject to which the field of immanence is only attributed by belonging to a self that necessarily represents such a subject to itself (reflection)."[65] Deleuze starts from the field of immanence and sees that the (human) Subject always remains above it thanks to the transcendental unity of apperception. I have started from the concept of the human (subject) and saw it as constituted – at least in its aspect of being different from the animal – by the reaching-out-into-transcendence that I have earlier identified as a clear mark of a plane of transcendence.

It bears repeating that Kant's plane of transcendence is explicitly theological: what the human (subject) reaches out to is God. Even though it might be a nonexistent God, he is nevertheless necessary for the human to be human. In Kant, transcendence is revealed as necessary for both scientific and ethical thinking.

A note on Levinas

The link between ethics and the plane of transcendence present in Kant is far more than an idiosyncrasy of the author of the three *Critiques*. While it is would lie far beyond the scope of this work to discern how prevalent this connection is in other manners of ethical thinking, it might be useful to present how the animal/human distinction operates in the philosophy of Emmanuel Levinas, whose thinking is at odds with Kant's on numerous points.

Levinas's philosophical project can be summed up as the endeavor to establish the primacy of ethics over ontology, which is especially interesting in the context of the preceding subchapter, where I have argued, among other things, that the distinction between man and animal in Kant, while often analyzed purely in terms of its ethical significance, can only be fully understood through the analysis of his ontology. Indeed, near the end of *Otherwise than Being* Levinas indicates that "Kantianism is the basis of philosophy if philosophy is ontology,"[66] a statement that Paul Davies calls the "harshest judgment" of Kant by Levinas.[67] Davies also compiles a simple comparison between the two thoughts: "1. Kant: respect (for the moral law); freedom; spontaneity; autonomy; 2. Levinas: responsibility (for the other); sincerity; passivity; separation; heteronomy,"[68] and while he then proceeds to problematize such a simplified juxtaposition of the two, showing possible areas where they are closer than usually thought, it remains clear that at least at face value (and according to Levinas' own words) rarely are two philosophical projects so far apart.

Levinas's conception of ethics of course quite explicitly relies on transcendence. In fact, one could argue that some of his most important texts are actually exercises in establishing an ethical plane of transcendence. For example, in the first chapter of *Alterity and Transcendence*, Levinas situates himself very consciously in the history of transcendence, on the one hand rejecting the simple idea of a "beyond," and on the other rediscovering (and, of course, any philosophical act of "discovery" is in fact conceptual creation) it in the face of the Other.[69] Similarly, the beginning of *Totality and Infinity* can be read as an endeavor of replacing the traditional metaphysical plane of transcendence with a new one, established with the help of the Cartesian idea of infinity.[70]

The transcendence-based ethical relationship is reserved for humans: "The dimension of the divine opens forth from the human face. A relation with the Transcendent [. . .] is a social relation."[71] Even if animals can elicit an ethical response in us, force us into the same passivity as humans can, it does not erase the primacy of the human face, which always serves as a model: "One cannot entirely refuse the face of an animal. It is via the face that one understands, for example, a dog. Yet the priority here is not found in the animal, but in the human face [. . .] The human face is completely different and only afterwards do we discover the face of an animal."[72]

In other words, there is no ethics worthy of its name without transcendence, and no relationship with an animal that could achieve transcendence; therefore, a relation with an animal may be *like* an ethical one only if the animal is thought of *like* a human – only through a mock-humanization of the animal can we achieve what will end up as a mock-ethical relationship with it. I will come back to this strategy in the next chapter, "False Immanence."

Notes

1　Indeed, it seems that no general book on animal studies written from a continental perspective can be complete without a chapter focusing on the German philosopher. Aside from the works cited later in this subchapter, as well as earlier cited analyzes by Derrida and Agamben, some of these include Matthew Calarco, *Zoographies: The Question of the Animal from Heidegger to Derrida* (New York: Columbia University Press, 2008), chapter I, Kelly Oliver, *Animal Lessons: How They Teach Us to Be Human* (New York: Columbia University Press, 2009), chapter 2, or Elizabeth De Fontenay, *Le silence de bêtes* (Paris: Fayard, 1998), chapter XVII, 3; shorter interventions include Philip Tonner's, "Are Animals Poor in the World? A Critique of Heidegger's Anthropocentrism," in *Anthropocentrism: Humans, Animals, Environments*, ed. Rob Boddice (Leiden: Brill, 2011), 203–21, or – from an Anglo-Saxon perspective – Paola Cavalieri, "A Missed Opportunity: Humanism, Anti-humanism and the Animal Question," in *Animal Subjects*, ed. Jodey Castricano (Waterloo, Ontario: Wilfrid Laurier University Press, 2008), 97–123.

2　There have even been attempts to argue that a Heideggerian approach can be useful for evolutionary biologists; e.g., Lawrence J. Hatab suggests that "evolution theory should find much more affinity with some of Heidegger's phenomenological concepts than with the discrete, atomistic, mechanical categories inherited from modern science." [Lawrence J. Hatab, "From Animal to Dasein: Heidegger and Evolutionary Biology," in *Heidegger on Science*, ed. Trish Glazebrook (Albany: State University of New York Press), 102.]

3　Martin Heidegger, *What Is Called Thinking*, trans. Fred D. Wieck and J. Glenn Gray (London, New York and Evanston: Harper and Row, 1968), 3.

4　Martin Heidegger, "Letter on 'Humanism'," in *Pathmarks*, ed. William McNeill (Cambridge: Cambridge University Press, 1998), 246–47.

5　Martin Heidegger, *The Fundamental Concepts of Metaphysics: World, Finitude, Solitude*, trans. William McNeill and Nicholas Walker (Bloomington and Indianapolis: Indiana University Press, 1995), 194.

6　Works concentrating on that subject include: Hubert Dreyfuss and Mark Rathall, eds., *Heidegger Reexamined, Vol. 4. Language and the Critique of Subjectivity* (New York: Routledge, 2002); Jeffrey Powell, ed., *Heidegger and Language* (Indianapolis: Indiana University Press, 2013); Krzysztof Ziarek, *Language After Heidegger* (Indianapolis: Indiana University Press, 2013) includes a scrutiny of recently published volumes of the philosopher's Collected Works (*Gesamtausgabe*) including important texts on the subject.

7　Heidegger, *The Fundamental Concepts of Metaphysics*, 177.

8　See Heidegger, *The Fundamental Concepts of Metaphysics*, 196–97.

9　"The stone cannot even be deprived of something like world." Heidegger, *The Fundamental Concepts of Metaphysics*, 196.

10　Heidegger, *The Fundamental Concepts of Metaphysics*, 197.

11　Heidegger, *The Fundamental Concepts of Metaphysics*, 199.

12　Heidegger speaks of *honey*, but of course a flower would contain nectar, and not honey. Brett Buchanan notes this and provides a discussion about how this apparent mistake changes the status of Heidegger's example [see Brett Buchanan, *Onto-Ethologies: The Animal Environments of Uexküll, Heidegger, Merleau-Ponty and Deleuze* (Albany: SUNY Press, 2008), 80–81].

13 Heidegger, *The Fundamental Concepts of Metaphysics*, 242.
14 Heidegger, *The Fundamental Concepts of Metaphysics*, 238.
15 Martin Heidegger, "On the Essence of Ground," in *Pathmarks*, 123.
16 Transcendence as something that governs the difference between man and animal in Heidegger's work is discussed by very few scholars, most notably Buchanan, *Onto-Ethologies*, 101 and ff. Death and the adage "The animal does not die" receives far more attention. However, Buchanan introduces the notion that man is transcendent, and, conversely, that the animal is immanent, only in passing, and does not make the connection between transcendence and death regarding the animal–human divide in Heidegger.
17 Martin Heidegger, *Being and Time*, trans. John Macquarrie and Edward Robinson (New York: Harper and Row, 1962), 290–91. Given here is, necessarily, a somewhat cursory analysis of the problem of death in Heidegger. For a more detailed analysis, see e.g., William D. Blattner, "The Concept of Death in Being and Time," in *Heidegger Reexamined, Vol. 1. Dasein, Authenticity and Death*, ed. Hubert Dreyfuss and Mark Rathall (New York: Routledge, 2002), 307–29; Adam Buben, *Meaning and Mortality in Kierkegaard and Heidegger* (Evanston: Northwestern University Press, 2016), 92–108.
18 Heidegger, *Being and Time*, 292.
19 Heidegger, *Being and Time*, 281–84.
20 Heidegger, *Being and Time*, 287–89.
21 See Heidegger, *Being and Time*, 289.
22 Heidegger, *Being and Time*, 299.
23 See, e.g., Martin Heidegger, "The Thing," in *Poetry, Language, Thought*, trans. Albert Hofstadter (New York: Harper Perennial, 2001), 161–84.
24 See Joachim L. Oberst, *Heidegger on Language and Death: The Intrinsic Connection in Human Existence* (London and New York: Continuum, 2009), 96–98.
25 D. Franck, "Being and the Living," in *Heidegger Reexamined, Vol. 1.*, 118.
26 See Hatab, "From Animal to Dasein," 96.
27 see Giorgio Agamben, *Infancy and History*, trans. Liz Heron (London: Verso, 1993), 52–53.
28 A quotation attributed to Benveniste (though without exact reference) by Agamben in *Infancy and History*, 54.
29 Calarco, *Zoographies*, 84; the original expression can be found in Georges Bataille, *Theory of Religion*, trans. Robert Hurley (New York: Zone, 1989), 23.
30 Agamben, *Infancy and History*, 56.
31 Agamben, *Infancy and History*, 57.
32 Agamben, *Infancy and History*, 57.
33 Agamben, *Infancy and History*, 58.
34 See Sigmund Freud, "Introductory Lectures on Psychoanalysis (part III)," in *Standard Edition of the Complete Psychological Works of Sigmund Freud, Volume XVI*, ed. James Strachey (London: The Hogarth Press, 1981), 284–85.
35 Sigmund Freud, "Future of an Illusion," in *The Freud Reader*, ed. Peter Gay (New York: W.W. Norton and Company, 1989), 686.
36 See Sigmund Freud, *Totem and Taboo*, trans. James Strachey (London and New York: Routledge, 2004), 164.
37 Freud, *Totem and Taboo*, 169–70.
38 The fullest account of the rise and fall of the Oedipus complex, updated to include the then-newly introduced terms of ego, id and superego, can be found in Sigmund Freud, "The Ego and the Id," in *The Freud Reader*, 628–60.
39 See, e.g., Szymon Wróbel, "Domesticating Animals: Description of a Certain Disturbance," in *The Animals in Us – We in Animals*, ed. Szymon Wróbel (Frankfurt am Main: Peter Lang, 2014), 219–38.

40 Sigmund Freud, "Group Psychology and the Analysis of the Ego," in *Standard Edition of the Complete Psychological Works of Sigmund Freud, Volume XVIII: Beyond the Pleasure Principle, Group Psychology and Other Works*, ed. James Strachey (London: The Hogarth Press, 1955), 116.

41 It might be interesting to note that Freud has a very romantic view of the Catholic Church, as he depicts it, and the relationships between those who form it, as those of a community of the faithful and not as primarily an institution of power.

42 Carl Schmitt, *The Concept of the Political: Expanded Edition*, trans. George Schwab (Chicago: University of Chicago Press, 2007), 25–26.

43 Carl Schmitt, *Political Theology: Four Chapters on the Concept of Sovereignty*, trans. George Schwab (Chicago: University of Chicago Press, 2005), 36.

44 Schmitt, *Political Theology*, 5.

45 Jacques Derrida, *The Beast and the Sovereign: Volume I*, trans. Geoffrey Bennington (Chicago: University of Chicago Press, 2009), 96.

46 Schmitt, *The Concept of the Political*, 55.

47 Schmitt, *The Concept of the Political*, 54.

48 Schmitt, *The Concept of the Political*, 36.

49 Jacques Derrida, *The Politics of Friendship*, trans. George Collins (London: Verso, 2005), 157–59.

50 The operative quotation here comes from Kant's *Lectures on Ethics* [trans. Louis Infield (New York: Harper and Row, 1963), 239.]: "So far as animals are concerned, we have no direct duties. Animals are not self-conscious and are there merely as means to an end. That end is man [. . .]. Our duties towards animals are merely indirect duties towards humanity. Animal nature has analogies to human nature, and by doing our duties to animals in respect of manifestations of human nature, we indirectly do our duty towards humanity."

51 Examples of a straightforward rejection of Kantian ethics with regards to animals can be found in Paola Cavalieri, *The Animal Question: Why Nonhuman Animals Deserve Human Rights*, trans. Catherine Woollard (New York: Oxford University Press, 2001), esp. 47–49 and Peter Singer, *Animal Liberation* (New York: Ecco, 2002). More generous readers of Kant include Tom Regan [see e.g. *Animal Rights, Human Wrongs* (Lanham: Rowman & Littlefield, 2003) or "The Case for Animal Rights," in *In Defense of Animals*, ed. Peter Singer (New York: Basil Blackwell, 1985), 13–26; Christine Korsgaard, "Fellow Creatures: Kantian Ethics and Our Duties to Animals," Tanner Lecture on Human Values delivered on February 6, 2004, accessed February 3, 2018, https://tannerlectures.utah.edu/_documents/a-to-z/k/korsgaard_2005.pdf or Martha Nussbaum, *Frontiers of Justice: Disability, Nationality, Species Membership* (Cambridge: The Belknap Press, 2007)], even though every one of these philosophers could certainly agree with Nussbaum, who admits that "Kantianism needs modification" (Nussbaum, *Frontiers of Justice*, 147).

52 For a detailed account and analysis of those, see: David L. Clark, "Kant's Aliens: The Anthropology and Its Others," *CR: The New Centennial Review* 1, no. 2 (Fall 2011): 201–89.

53 Immanuel Kant, "Anthropology from a Pragmatic Point of View," in *Anthropology, History, and Education*, trans. Robert B. Louden (Cambridge: Cambridge University Press, 2007), 239.

54 Kant: "[the unity of apperception] is [the] transcendental ground for the unity of the consciousness in the synthesis of the manifold in all our intuitions, hence also of the concepts of objects in general, consequently also of all objects of experience, without which it would be impossible to think of any objects for our intuitions." Immanuel Kant, *Critique of Pure Reason*, trans. and ed. P. Guyer and Allen W. Wood (Cambridge: Cambridge University Press, 1998), 232.

55 Kant, *Critique of Pure Reason*, 238.

56 Kant, *Critique of Pure Reason*, 240.
57 In the *Critique of the Power of Judgment*, Kant shows that we can and indeed do see nature as working with regard to a final cause (an end); however, this has little to do with empirical (scientific) understanding of nature.
58 Kant, *Critique of Pure Reason*, 577.
59 "Thus, in the union of pure speculative with pure practical reason in one cognition, the latter has primacy, assuming that this union is not contingent and discretionary but based a priori on reason itself and therefore necessary." Immanuel Kant, "Critique of Practical Reason," in *Practical Philosophy*, trans. and ed. Mary J. Gregor (Cambridge: Cambridge University Press, 1999), 237.
60 Kant, *Critique of Pure Reason*, 675.
61 Kant, *Critique of Pure Reason*, 533.
62 Immanuel Kant, "Groundwork for the Metaphysics of Morals," in *Practical Philosophy*, 37–109.
63 Immanuel Kant, *Critique of the Power of Judgment*, trans. Paul Guyer and Eric Matthews (Cambridge: Cambridge University Press, 2000), 316.
64 Kant, *Critique of the Power of Judgment*, 316, footnote.
65 Gilles Deleuze and Felix Guattari, *What Is Philosophy?* trans. Hugh Tomlinson and Graham Burchell (New York: Columbia University Press, 1994), 46.
66 Emmanuel Levinas, *Otherwise Than Being*, trans. Alphonso Lingis (Dordrecht: Kluwer, 1991), 179.
67 Paul Davies, "Sincerity and the End of Theodicy: Three Remarks on Levinas and Kant," in *The Cambridge Companion to Levinas*, ed. Simon Critchley and Robert Bernasconi (Cambridge: Cambridge University Press, 2004), 164.
68 Davies, "Sincerity . . .," 167.
69 Emmanuel Levinas, "The Other Transcendence," in *Alterity and Transcendence*, trans. Michael B. Smith (London: The Athlone Press, 1999), 1–76.
70 See Emmanuel Levinas, *Totality and Infinity*, trans. Alphonso Lingis (The Hague: Martinus Nijhoff, 1979), esp. 33–52. For a more complete account on Levinas' view of transcendence, especially in relation to ethics, see, e.g., Robert Bernasconi, "No Exit – Levinas' Aporetic Account of Transcendence," *Research in Phenomenology* 35 (2005): 101–17. See also Jean-Francois Rey, *La mesure de l'homme: L'idée de l'humanité dans la philosophie d'Emmanuel Levinas* (Paris: Michalon, 2001), esp. 89–91, where Rey analyzes how the relation to the human face creates a reference to the dimension of height/verticality (*hauteur*).
71 Levinas, *Totality and Infinity*, 78.
72 Tamra Wright, Peter Hughes, and Alison Ainley, "The Paradox of Morality: An Interview with Emmanuel Levinas," in *The Provocation of Levinas. Rethinking the Other*, ed. Robert Bernasconi and David Wood (London and New York: Routledge, 1988), 169–72.

5   False immanence

Why philosophy presents a negative vision of the animal

The previous chapter showed that each of the classical definitions of man is based on thinking in terms of a plane of transcendence. What is proper to man – be it language, politics, reason or ethics – requires him to function in a way that allows him to reach out or upwards for an absolute point of reference, and only thanks to this point of reference could he be essentially human. Of course, this "essence" is always understood in a slightly different way, not necessarily corresponding to what is called "essence" in classical philosophy, yet the essential character of reaching out to a stable point of reference remains visible.

In the case of Heidegger's language-based definition, the point of reference was death – saying that animals do not have language amounts to the fact that animals do not die, or at least do not die "properly," only perish. In order to have language, or rather to be in the way of having language, man needs to transcend the order of things, and he can only do so thanks to his ability to have a certain relationship with death.

Strangely enough, Agamben's theory of infancy makes exactly the same point, although it seems to start from a radically different assumption – man is the only being who does not have language (at a certain point of his development), and can therefore exist both in the semiotic and the semantic realm of language. Existence in the semantic realm – human existence – is only possible thanks to the ability to transcend the semiotic immanence of nature.

In Freud's work, the life of man as a political animal is based on a plane of transcendence at least in three aspects – phylogenetically, for the establishment of culture or the *polis* it was necessary to form a link with the murder of the father; ontogenetically, a transcendent to the ego (though "internal") point of reference needs to be created in order to participate in culture; finally, the forming of groups requires a reference to a common person or idea. A similar pattern, although purely political, can be seen in Schmitt's theory. The political requires constructing a plane of transcendence with the sovereign and the enemy serving as external points of reference; neither of these acts of construction is available to the animal.

In Kant's scheme, theoretical reason needs to structurally think as if there were a supreme, necessary being – God. The same goes for practical reason: in order to think practically, to produce universal moral maxims, practical reason needs a regulative principle, to which it can anchor itself in order to ensure the universal character of the maxims – and this principle again is God. Kant's God is transcendent in a very strong sense – he exists beyond the realm of all possible experience. The plane of transcendence proves thus to be a theological plane, as Deleuze and Guattari theorized. In fact, and this will be crucial in this chapter, God's existence is irrelevant to the functioning of reason in its practical or theoretical aspect alike. What is important is the structure of thinking, which is based on a classical plane of transcendence.

When moving to the realm of practical reason, Kant also moved to another definition of man – he is the only creature that can be truly moral; here, too, the structure of thinking is more important for my argument than its content. Levinas' theory, even though aimed directly against Kant (and Heidegger, for that matter), reaches the same conclusion: the ultimately human trait, that is, the ability to be ethical, is based on a relationship with transcendence.

What do these conclusions mean for thinking the animal? Since a man is human only insofar as he "transcends," and – as we saw in "Anatomy" – the animal is what is left of man when the properly human is taken away, then the animal is "immanent." This conclusion was reached explicitly in the case of Heidegger, but the same can also be said about other philosophers mentioned in the last chapter. Agamben claims the animal functions only in the semiotic realm of language in which it swims as "water in water." According to Kant, the animal only exists in the immanent realm of immediate experience, cannot escape its immediateness, and only reacts in a pathological way, unable to judge its behavior from an external standpoint. With Freud, the animal cannot build a superego (on the individual and group level) that functions as a built-in reference to the laws thanks to which culture as the voluntary suppression of some drives is possible.

What is more, a similar conclusion can be reached based on other writers mentioned in this work, albeit in a different context – for example, Descartes' "reacting" animal cannot "respond" because it cannot transcend its immediate experience, and Lacan's animal cannot transcend its instinct and therefore is not capable of feigning a feint.

This logical conclusion is too simple not to raise any suspicion. It is almost an analytical statement – if the human is transcendent and the animal is not human, then the animal is not transcendent, therefore the animal is immanent. However, the thesis about the transcendence of the human and the immanence of the animal has far-reaching consequences to the relationship between the two, and helps elucidate the way that the concepts "animal" and "human" are constructed.

Classically, the relationship between transcendence and immanence is explained thusly: first, there is the realm of immanence: this is "our world" or (as Kant would have it) "possible experience." Only after there is transcendence, an "outer world," where God resides. Immanence precedes transcendence – in the order of discovery, first man sees himself and the world around him and then infers the existence of an "outer world." First, he is animal, and only later, by virtue of his ability to reach transcendence, he becomes something else.

However, in the case of the animal and the human, the opposite seems to be the case. With each of the analyzed philosophers, we seem to come across a similar pattern, even if it is not consistent in a chronological manner or the manner of exposition: first, man is defined as human according to the plane of transcendence, and only then is the animal defined as missing an important trait, so that it cannot be human. Agamben needs to define language as semiotic and semantic in order to deprive the animal of the latter – the animal has (and has always had) language, but only an untrue, distorted version of it. Kant first notices the role of practical reason, and then deprives the animal of it – it has experience, but without reason, it is unable to react in any way that is not mechanical or pathological.

The case of Heidegger is particularly interesting, as seemingly the thread of his thinking is going in the other direction – starting from the worldless stone, continuing through the poor-in-world animal and finishing with the world-forming man. What is more, he develops the theory of the disinhibiting ring to explain animal behavior and dedicates almost 90 pages to the analysis of the thesis that the animal is poor-in-world – about the same amount of time he devotes to the world-forming man. And yet, even – or maybe especially – in Heidegger's text, the aforementioned tendency is also present. The organizing principle which lies behind the whole second part of his text is "What is world?" – and, as I have shown in "Anthropology," it is only man who truly has world; it is only Dasein that is world-forming. This means that all the developments on the animal are not written from a perspective of purely scientific interest – or rather *desinteressement* – but instead form an important part of the argument for the uniqueness of man. The real question behind Heidegger's analysis of animal behavior can be posed as this: how to explain it without recourse to language, without letting them have language, without allowing them this human prerogative. Heidegger's reasoning seems to be driven by a tacit axiom – *given that they do not have language*, how can we explain animal behavior? It is not by accident that he chose bees for his analyzes they are one of philosophy's most problematic species. By explaining away the possibility of bees being *zooi politikoi*, Heidegger solidifies the human–animal caesura.

The tendency to explain animal behavior using such an axiom, or to explain it negatively, is in no way a new one, and is not limited to the philosophers I analyzed in the preceding two chapters. In fact, it seems to be a philosophical *leitmotiv* from the times of Descartes on. Let me once again refer to his letter to the Marquess of Newcastle:

> there has never been known an animal so perfect as to use a sign to make other animals understand something which bore no relation to its passions; and there is no human so imperfect as not to do so.[1]
>
> I know that animals do many things better than we do, but this does not surprise me. It can even be used to prove that they act naturally and mechanically, like a clock which tells time better than our judgment does. Doubtless when the swallows come in spring, they operate like clocks. The actions of honeybees are of the same nature; so also is the discipline of cranes in flight; and of apes in fighting, if it is true that they keep discipline.[2]

Descartes' obstinacy is almost amusing. If animals act in a way inferior to men – for example fail to respond and simply react – it is proof that they are merely machines, deprived of souls. However, if they act in superior ways – are better disciplined like cranes or apes; function in complicated societies as honeybees do – it proves exactly the same thing. Descartes – and with him most philosophers – keep repeating "you are a machine, you are a machine" to anything an animal does, with the obstinacy of a stereotypical psychoanalyst yelling "it is your mother" at their patient.

In this view, philosophers and animals form a tandem much akin to Achilles and the tortoise – only this time it is the animal that does the chasing. Any time an animal does something that has been hitherto thought as proper to man – speaks, reasons, cheats, etc. – the philosopher seems to already have taken a minuscule step forward, saying "but that is not *really* speech or reasoning or feigning, because the *real* thing is something more." It is not enough to feign, says Lacan (or Dennett, for that matter);[3] you must also feign feints. It is not enough to relate to things in the world, says Heidegger; you must also *have* world, and this is completely different.

This kind of thinking about man and animal is best visible in cases where this axiomatic is dogmatically or methodologically dropped. This is the case with some strains of sociobiology or evolutionary biology. In the same way that Descartes explains the perfect nature of cranes in flight by their mechanical behavior and Heidegger meticulously tries to prove that bees are only driven by instinct in their seemingly goal-driven behavior, Dawkins will prove the instinctive nature of unselfish deeds. The question here again contains a tacit axiom: *provided that anything can be explained without recourse to traits treated as essentially human,* how can we explain complex human behaviors? This is not Descartes's theory backfiring, but its logical continuation.

Before moving forward, let me sum up the current developments: the pair of concepts "the human" and "the animal" has a clear hierarchy. The human is the first one to be established, and the animal is always secondary; it always follows the form "animal = man minus essentially human trait(s)." As I established in the chapter "Anatomy," it can be said that rather than a rational (political, ethical) animal, man is animal *and* rational, animal *and* political and so on. This equation is – still following Agamben – actually a subtraction: the human is established by subtracting the animal from man. The animal, in turn – as we can assert in light of what has been said in this chapter – is man *minus* rationality, man *minus* politics, man *minus* human.

The two preceding chapters have shown that the human is a concept created on a plane of transcendence, always needing an absolute point of reference that lies beyond the reach of the only "immanent" animal. But, since the concept of the human is clearly first in this pair and animal is only a derivation, the seemingly natural order – from immanence to transcendence – is turned on its head. When one starts from transcendence, immanence will always be subjugated to it, it will always be viewed from the already-established plane of transcendence, and therefore will always be twisted and impure, stained by the shadow of what looms above it. Each theory of transcendence includes a theory of immanence, but only as its secondary aspect, which cannot be taken seriously on its own.

What philosophers have been labeling "animal," the concept of what is not human in man, is exactly this – misunderstood immanence. It is not surprising that the animal is usually depicted in a negative way, as deprived (of world, reason, language), and yet somehow perfect, even in a limited, mechanical way. And while such conceptual impoverishment is perfectly understandable within philosophies that have the establishment of a transcendence-based concept of the human as their explicit or implicit goal, it is also eerily present in so-called "animal philosophy." This way, instead of constructing a positive concept of the animal, based on a plane of immanence, these philosophies endeavor to find a place for animals in the old framework, leaving it intact and, in fact, conserving it. The result is only – to put it in Agamben's terms – moving the caesura between the animal and the human, without questioning the logic of transcendence because of which there is a caesura in the first place.

This result is usually accomplished through the use of one of two strategies, which could be called overanimalization and overhumanization. Overanimalization occurs when the caesura is moved "downward," towards the animal, while overhumanization means moving the caesura "upward," towards the human. Put simply, the first strategy means claiming "humans are animal," and the second "animals are human."

Dawkins' theory of the selfish gene, that I have glossed over previously, is a good example of the first strategy. In the chapter *Zoology* I mentioned what Matthew Calarco calls the "identity" approach – any evolutionary theorist would have to go to great lengths not to subscribe to one or another version of it. For example, it is embraced by Robert Trivers in the foreword to Dawkins' *opus magnum*:

> The chimpanzee and the human share about 99.5 per cent of their evolutionary history, yet most human thinkers regard the chimp as a malformed, irrelevant oddity while seeing themselves as steppingstones to the Almighty. To an evolutionist this cannot be so.[4]

Dawkins soon follows suit: "We are survival machines, but 'we' does not mean just people. It embraces all animals, plants, bacteria, and viruses."[5] A "survival machine" is a vessel thanks to which genes can survive; it is a carrier that can be used for protection and reproduction of the gene, which is the basic brick of Dawkins's theory. Genes are the true individuals in his ontology, and from the point of view of the gene, there is not much difference between a virus or a human being, provided that it helps to survive and reproduce – in this way genes are "selfish," for the survival of an individual virus, plant or animal is only a means to their own survival. At first it seems promising that Dawkins's theory allows a common denominator for humans and animals, even plants and viruses, or even beings that are not alive at all. He begins his story with a description of how single molecules "learned" to reproduce, which is the first step on the long road to animal or human life and genes. The details are unimportant for the point I am trying to make, so I will only cite the key ending fragment:

> [T]he story of the replicator molecules probably happened something like the way I am telling it, regardless of whether we choose to call them "living." Human suffering has been caused because too many of us cannot grasp that words are only tools for our use, and that the mere presence in the dictionary of a word like "living" does not mean it necessarily has to refer to something definite in the real world. Whether we call the early replicators living or not, they were the ancestors of life; they were our founding fathers.[6]

By telling the story in this way and refusing to draw a definite line between living and not-living beings, not to speak of animals and humans, Dawkins goes a long way towards "renaturalization," a strategy which is key for thinking in terms of immanence. Even so, the theory of the *Selfish Gene* fails to speak of the animal in terms of true immanence.

I have already said that Dawkins's approach is an example of how Descartes's meticulous explanations of seemingly purposeful actions in terms of strict causality can backfire. It is not only animal behavior that can be explained this way – human as well. This is exactly what Dawkins does, and his conceptual framework (even if it is metaphorical), remains highly Cartesian: "we animals are the most complicated and perfectly designed pieces of machinery in the known universe."[7]

Of course, referring to animals as "machines" is a typically Cartesian move. What Descartes does is, as we have seen in Chapter 1, to separate the human (soul-transcendence) from the animal (machine-immanence) within man, and then to understand all "animals" in terms of machines. From what I said earlier in this chapter, it should be clear that this clearly is a movement of establishing a machine as a false immanence, only to be understood negatively in the face of the positive transcendence. Dawkins (and his name here is just a metonymy for all similar theories)[8] only takes the "machine" part of Cartesianism, which will always remain a crippled, false immanence. Every use of terms such as machine, instinct, etc. will leave us thinking in terms of a plane of transcendence. This way, the "survival machines," men and animals alike, are naturally understood as blind vessels for (equally blind) genes.

Naming Dawkins as one of the proponents of the first strategy is not meant to suggest that it can be found solely among biologists, nor that the second strategy is used predominantly by, for example, "compensatory humanists" (as Rosi Braidotti calls liberal thinkers who try to open classical humanities to the animal question). For example, while introducing Kant's take on the *animal rationale* definition of man, Christine Korsgaard shows a possible reinterpretation of the philosopher's claims, which could build an ontology upon which a more inclusive ethics can be superimposed. Korsgaard – who does not wholeheartedly agree with the argument herself – shows how animals can be treated as ends in themselves, since it is our own animal nature (and not just our legislative will, as Kant would have it) that makes us treat *ourselves* as ends in themselves.[9] This move allows her to circumvent Kantian exceptionalism, but it comes at a cost – the ethical community between man and animal is built by underscoring the animal in man, however without the elimination of the human/animal caesura and the crippled immanence theory that comes with it.

A more complicated (and implicit) relationship between immanence and transcendence – between man and animal – can be found in the so-called "rights" approaches in animal ethics, especially in the most abolitionist visions present in the writings of Tom Regan, Gary Francione, Patricia MacCormack and others. While a radical immanence approach that I will be presenting in the next chapters offers a practical critique of radically abolitionist approaches (I will present it in the last chapter), right now I would like to focus on the theory they presuppose.

Rights approaches generally start from a Kantian premise that animals have "intrinsic value" or "dignity" similar to that which forms the basis of "human rights" in their various formulations. Of course – in a move similar to Korsgaard's argumentation that I cited previously – this value needs to be based on something less rigorous then a (super)human rationality of the Kantian kind. For Regan, the key distinction is that some beings (including, but not necessarily limited to, mammals, birds and fish) can be said to be "subjects of a life"[10] – in other words, they are not mindless machines, but their life matters to them.

In the case of the philosophers I mentioned previously, the rights stance is accompanied by the abolitionist stance – the radical elimination of all the ways in which the intrinsic dignity of animals is abused, including at least commercial farming, scientific experiments, captivity and obviously all forms of hunting.[11]. However, the – doubtlessly noble – moral absolutism of abolitionist approaches rises a problematic question of what interactions (if any) humans can have with animals in order to leave their dignity intact. Let me give a few examples.

Sue Donaldson and Will Kymlicka – who are not strictly abolitionists in a radical sense of the term – in their remarkable *Zoopolis* present a political theory of animal rights that divides animals into three categories, each one modeled after a category already present in "human" politics. In this vein, domesticated animals who live alongside us would be "citizens" and therefore have three basic rights: "residency (this is their home, they belong here), inclusion in the sovereign people (their interests count in determining the public good), and agency (they should be able to shape the rules of cooperation)."[12] "Liminal" animals, those who live in cities or other human dwellings, but are not domesticated and generally shy away from contact, should be given the slightly looser status of denizens.[13] Lastly, wild animals should be likened to foreigners and granted sovereignty on their own land, and humans should respect that sovereignty, which means "protecting the freedom, autonomy, and flourishing of wild animals, and in general [. . .] [being] very cautious about intervening in nature."[14] This last category of animals is the most interesting for the case at hand, as (even with the caveats Donaldson and Kymlicka make) it shows that what is desirable is a caesura between humans and animals which is as strict as possible, with our duties to other types of animals shaped exactly by the failure to respect this caesura – for example, domestic animals should indeed be citizens, because we have "brought such animals into our society, and deprived them of other possible forms of existence (at least for the foreseeable future)";[15] in other words, it seems we are to hold ourselves responsible for the crossing of barriers that should have been left intact.

This point is visible more clearly in some of Gary Francione's writing. Francione – while obviously championing giving the best possible care to already existing domestic animals – says that "if the world were more just and fair, there would be no pets at all."[16] In the history of domestication, we are guilty of subordinating animals to ourselves, making them unable to live without our help, and indeed encouraging their helplessness. None of this would be acceptable if it were happening to another human being; it should not be happening to animals if we claim, as proponents of rights theories do, that animals have equal dignity or intrinsic worth. Eliminating all domesticated animals (of course, with proper care given to actually existing domesticated animals) – in a way reversing domestication – is the only way to ensure that no rights are violated.

This vein of abolitionist thinking is followed to a radical outcome in the writings of "new abolitionist" Patricia MacCormack, who argues that the only way for true elimination of animal suffering and other malicious effects of human–animal interaction is eliminating the human altogether. Proposing a stance of voluntary extinction, MacCormack says that "[a]bolitionist ahuman ethics are only truly possible if we are not here".[17] As she comes from a background that differs from the Kantian inspiration of most rights-based abolitionisms, MacCormack is perhaps better able to present the actual stakes of such abolitionism, which are not so much physical – as in the mechanical separation of animals from human beings – but metaphysical. As she points out: "Humans continue to show their want for interaction with non-humans. Want is monodirectional; love is gracious acknowledgement of the relations we have made and those we must inevitably continue to make. Love will not cease at human absence."[18] In other words, true and reciprocal relationships between human beings and animals will only be possible after the former cease to be human and turn into something else entirely. That the only path to such a ceasing is the extinction of humans is as accidental as it is practically necessary. To use the language of this work, I might say that it is the "human" in man that stands between a truly just relationship (what MacCormack calls "love") between human beings and animals; and, as I have been trying to show in the preceding chapters, this "human" in man is precisely transcendence.

It seems that – on a theoretical level – the actual reason for minimizing animal–human contact is already present in the first assumption of the rights approach, although it most probably expresses something unthought in the movement. Admitting that we can speak of animal "dignity" or "intrinsic value," in the same way that we can speak of such value in the human, implicitly means granting the animal the right to transcendence, or at least to something which was usually grounded in transcendence. But this "humanization" of the animal is not complete – while the animal ceases to be completely "immanent" (within the scope of purely negative immanence), it does not achieve the full capabilities of transcendence; especially, it is not judged as capable of doing evil (it can feign, but it cannot lie; it can kill, but it cannot murder, etc.). All this possibility to do evil – as well as judge good and evil – is left to "man." To use Agamben's terms again (and to perhaps over-exaggerate a bit), the caesura between man and animal has again been moved so that man is produced as someone whose malignant nature taints everything he touches. The only way to remove the evil of man is to remove the human – either by creating physical barriers or by extinction, thus leaving behind an animal that is equally "natural" (with all the naively romantic connotations of the term) as it is "humanized."

Such examples of "false immanence" help to establish the place that philosophy now holds concerning the thinking of the animal. We are in a place structurally similar to what Nietzsche said about the death of God. Knowledge that God is dead, which labels itself "atheism," is common; it is taken as obvious in any one of Nietzsche's descriptions – the first reaction of the people who hear the news is disbelief or laughter. However, the problem is we do *not* know exactly what it means – the "factual" nonexistence of God does not result in his "structural" nonexistence; even if we *know* that God is not there, we keep his place empty or filled with more or less absolute referents who serve his place in our thinking.

When it comes to thinking the animal, we also *know* there is no such thing as the animal as represented in the philosophical concept I have been trying to understand here, animal as a radically different being than man, a concept that refers to chimpanzees as well as ants and that does not refer to the human. We *know* that virtually any difference between animals and humans is a quantitative, not a qualitative, one. And we know that we were not created by a transcendent God as well as we *know* that we do not need God to think or be good people (or animals). However – and it is the first important conclusion from the analysis of the concept of the animal – we are still pious. We still think *as if* there was such a radical difference – and even if we situate it differently, judging that apes are worthy of our moral consideration while, say, shrimp are not, then we still think in terms of the same plane that was behind establishing the difference in the first place. The realization that animals are worthy of being accepted in our moral space is in fact saying that "animals are people, too" – or, in the most radical cases of the abolitionist approach, that they are in fact better at being people than we are; they are as people should be.

While this first conclusion might rightfully be called a "negative" one, the second one is more positive. It is not by accident that it was the concept of the animal that lead us to the problem of immanence in the first place. This concept and the problems that arise when philosophers try to use it and so meticulously separate the human from the animal are the first indicators as to how one should think in terms of immanence. They indicate the core problems that need to be addressed in order for us to be able to think in such a manner – to once again have recourse to Deleuzian terminology, the animal helps form a line of flight towards the thinking of immanence. The next chapter will be devoted to exploring several such areas and key problems, namely that of thinking, ethics, politics and death.

A note on Yovel

Before proceeding to the positive elaborations, I would like to analyze one more theory – this time an explicit theory of immanence – that will further allow to identify the core problems that stand before such a theory.

In the second volume of his work on Spinoza, entitled *The Adventures of Immanence*,[19] Yirmiyahu Yovel discusses different philosophies of immanence, including those of Kant, Feuerbach, Marx and Freud. Here I will focus on the last chapter, where he seeks to establish guidelines for his own theory of immanence.

Inspired by Kant, Yovel distinguishes two types of theories of immanence – dogmatic and critical. The first kind treats immanence as an infinite absolute in a move that might be called deification (Spinoza's God=Nature as *causa sui* fits this description perfectly, but so does the Marxist this-worldly salvation idea). The critical type needs to take into account the basic finitude of human exist-ence, which means, among other things, the realization that we can only interpret our "world" in a fragmentary manner, and that no set of values or norms can be deemed universally valid.[20]

Therefore, "A philosophy of immanence must place all sources of normative-ness in the actual world."[21] How does Yovel envisage this source of normativeness?

> Man [. . .] existing as self-transcendence within immanence, creates mean-ings for himself and the world around him; it is through his own works, desires, projects and endeavors that morally significant meanings emerge in the world. No external, timeless guidance is available to him, whether he wel-comes it as salvation or rejects as bondage. And since human life is neither static nor repetitive, the ethical universe, too, assumes various faces and is open to change.[22]

Upholding the idea of self-transcendence of man makes Yovel a critical Feuer-bachian, and shows that, in fact, he remains within what I have been identifying as a plane of transcendence. Indeed, he does not believe in the absolute essence of Man, as Feuerbach did, but opts for an interchanging version of it. What Yovel does is to move transcendence within (much like the author of *The Essence of Christianity*), and then to treat it as an empty slot to fill in with various values. Though the model becomes dynamic, its structure stays the same.

It is easy to see why Yovel would reach such a conclusion, if one looks again at the previously cited fragment – Yovel treats human beings as principally *rational*, and this rationality is of a very Kantian kind. Paired with finitude, rationality of this type is a crucial obstacle – from the point of view of this work – to achieving true immanence. In the words of Yovel, who nevertheless treats this as a "most delicate or controversial point":[23]

> [A]s finite beings we can neither affirm the transcendent domain nor rid our-selves completely of its empty yet meaningful horizon. By "empty" I mean

that it cannot be filled with any positive contents or even be asserted to exist. Yet this empty horizon is meaningful as a memento of our own finitude and a critical barrier against turning the immanent world into an absolute or a kind of God.[24]

And further:

It must be stressed that concern with transcendent issues is a *rational* necessity according to Kant and an authentic human stance. [. . .] Critical rationality requires maintaining the mind's transcendent quest along with the recognition that it cannot be fulfilled. It thereby creates an unfulfilled gap, a void, a tension, which is the mark of human finitude and a distinctive feature of the critical philosopher.

Thus critical rationality can neither abolish the perspective of transcendence nor give it concrete content and substance. Transcendence hovers over the immanent domain as a question mark, a possibility that will always remain empty for us. Since transcendence, on this critical view, is neither an entity nor a straightforward fiction, we may refer to it as a "horizon" which our finitude projects over being but which we may not solidify or fill with any objects. Nor can we adequately talk about it except in a roundabout way, by metaphor which points to this horizon without fixing anything actual about or even within it.[25]

It seems that Yovel's key problem lies in the very beginning of the second cited fragment – indeed, concern with transcendence is both *rational* and *human*, because, as we have seen in the preceding chapters, it is precisely the human and the rational that is defined by transcendence. It seems therefore that, Yovel's theory of "immanence" is a rather classical example of a plane of transcendence, since he cannot think otherwise than by referring to the transcendent domain, even if the latter remains just a question mark. If a theory of immanence is to be strict, it must resign from this understanding of rationality. In fact, it must also resign from traditional "human" interest in transcendence – it might be "natural" for the human to be interested in it, but the question is: can we do without it, can we think otherwise?

However, Yovel's theory is not without merit for the project I am trying to accomplish in this book. It delineates two key points of interest for a theory of immanence – on the one hand, and following Kant, it shows the importance of a new approach to what was called "rationality" and its reliance on transcendence; on the other hand, and in a way following Heidegger, it underlines the key function of finitude in discussing the plane of transcendence. It also stresses the necessity of a *critical* stance when it comes to establishing a plane of immanence – immanence cannot be treated as an absolute, for indeed it would simply restore the transcendence it planned to get rid of in the first place. All those points will need to be taken into account in the final parts of this study.

Notes

1 René Descartes, *The Philosophical Writings of Descartes, Vol III: The Correspondence*, trans. John Cottingham, et al. (Cambridge: Cambridge University Press, 1991), 303.
2 Descartes, *The Correspondence*, 304.
3 For the comparison between Dennett's and Lacan's arguments see Cary Wolfe, "Thinking Other-Wise, Cognitive Science, Deconstruction and the (Non)Speaking (Non) Human Subject," in *Animal Subjects*, ed. Jodey Castricano (Waterloo, Ontario: Wilfrid Laurier University Press, 2008), 125–43.
4 Robert L. Trivers, Foreword to Richard Dawkins, *The Selfish Gene* (Oxford: Oxford University Press, 2006), xix.
5 Dawkins, *The Selfish Gene*, 21.
6 Dawkins, *The Selfish Gene*, 21.
7 Dawkins, *The Selfish Gene*, xxi.
8 For a summary of how Dawkins's theory is related to classical deterministic schemes, see, e.g., Rob Preece, "Selfish Genes, Sociobiology and Animal Respect," in *Animal Subjects*, 39–62.
9 Christine Korsgaard, "Fellow Creatures: Kantian Ethics and Our Duties to Animals," Tanner Lecture on Human Values delivered on February 6, 2004, 104, accessed February 3, 2018, https://tannerlectures.utah.edu/_documents/a-to-z/k/korsgaard_2005.pdf.
10 See Tom Regan, "The Case for Animal Rights," in *In Defense of Animals*, ed. Peter Singer (New York: Basil Blackwell, 1985), 13–26.
11 See Regan, "The Case for Animal Rights."
12 Sue Donaldson and Will Kymlicka, *Zoopolis: A Political Theory of Animal Rights* (Oxford: Oxford University Press, 2011), 101.
13 Donaldson and Kymlicka, *Zoopolis*, 210–51.
14 Donaldson and Kymlicka, *Zoopolis*, 186.
15 Donaldson and Kymlicka, *Zoopolis*, 101.
16 Gary L. Francione and Anna E. Charlton, "The Case Against Pets," *Aeon*, September 8, 2016, https://aeon.co/essays/why-keeping-a-pet-is-fundamentally-unethical.
17 Patricia MacCormack, "After Life," in *The Animal Catalyst: Towards Ahuman Theory*, ed. Patricia MacCormack (London: Bloomsbury, 2014), 183.
18 MacCormack, "After Life," 187.
19 Yirmiyahu Yovel, *Spinoza and Other Heretics: The Adventures of Immanence* (Princeton: Princeton University Press, 1989).
20 Yovel, *Spinoza and other Heretics*, 175–76.
21 Yovel, *Spinoza and other Heretics*, 177.
22 Yovel, *Spinoza and other Heretics*, 177.
23 Yovel, *Spinoza and other Heretics*, 173.
24 Yovel, *Spinoza and other Heretics*, 173.
25 Yovel, *Spinoza and other Heretics*, 174.

6 Radical immanence

Recreating the concept of the animal

The role of the previous chapter was primarily negative – it focused on identifying false theories of immanence, theories that, although they seem to be speaking about pure immanence, are in fact trying to comprehend it from a point of view of the plane of transcendence. In these philosophies, immanence is viewed as a negative, as something mechanical or instinctive. Probably the most important point made in that chapter was that it is not enough to "physically eliminate" the transcendent referent in order to think in terms of a plane of immanence – indeed, this is the worst way to go about it, for it makes thinking itself impossible, for thinking, as understood in terms of the plane of transcendence, requires such a referent to "work." Secondly, as we have already seen in Kant, the existence of the transcendent referent, and even its physical position is irrelevant – what is relevant is the structure on which the creation of concepts occurs.

In this chapter, I will try to provide some groundwork for a theory of immanence that would meet the high standards that have been outlined in the preceding one. The simplest way to understand these standards would be to speak of "radical immanence," in parallel to what Martin Hägglund calls "radical atheism."[1] In his words:

> Atheism has traditionally limited itself to denying the existence of God and immortality, without questioning the desire for God and immortality. Thus, in traditional atheism mortal being is still conceived as a lack of being that we desire to transcend. In contrast, by developing the logic of radical atheism, I argue that the so-called desire for immortality dissimulates a desire for survival that precedes it and contradicts it from within.[2]

Two points are important here. Firstly, radical atheism denies the desire for transcendence – it says that not only there is no God (we saw in the Nietzsche sections of the preceding chapter that this is in fact a banality), but that we should not miss his existence, we should not try to fill the void with something else. This is also the core of what I understand as radical immanence – immanence without any transcendence whatsoever, without it as a desired, expected, presupposed object, nor as an explosion from within. The second part of Hägglund's notion is more problematic – at first glance, it looks like he was trying to find hidden immanence (desire for this-worldly survival) behind transcendence (desire for God and immortality of the soul). His stance, however, is not that simple, and he is not a philosopher of immanence whatsoever. This is primarily because his theory is strongly based on the Derridian notion of auto-immunity of life.[3] While this is not a place to go too deeply into detail, let me just point out that the basis for this notion is that life contains in itself a principle of death – the key concept of survival in Hägglund's theory is precisely the clash of the two principles:

> The notion of survival that I develop is incompatible with immortality, since it defines life as essentially mortal and as inherently divided by time. To survive is never to be absolutely present; it is to remain after a past that is no longer and to keep the memory of this past for a future that is not yet. I argue that every moment of life is a matter of survival.[4]

This way of defining life, supposing its essential link with death, is profoundly Heideggerian – not surprisingly for Derrida – and we have already seen in the chapter devoted to the German thinker that death in fact serves as the absolute referent that is key to a plane of transcendence. I will get back to this problem when discussing Braidotti's understanding of death towards the end of this chapter.

For most of this text I have endeavored to establish a link between the concept of the animal and immanence. Chapter 4, devoted primarily to Heidegger and Kant, helped to show that the classical definitions of man rely on transcendence, leaving immanence (a secondary notion in such a plane) to the animal realm, first and foremost within man, and only secondly outside him. This immanence – and, *a fortiori*, the animal – was understood in a negative, partial way, only serving as a springboard for the human to reach transcendence. This strategy led to defining the animal primarily negatively – it does not think, does not act ethically, does not form political entities, does not die, *but only* reacts instinctively or pathologically, forms herds, perishes, etc.

However, the animal is crucial for the study of immanence, for – as falsely as it was viewed beforehand – it remains its primary settlement. This is why the efforts to delineate primarily "human" traits and structures willy-nilly also lead to delineating important areas in which a radical theory of immanence can bloom. For example, the phrase "the animal does not die" actually means "death is incomprehensible without recourse to transcendence." It goes without saying that this means simply that the human is incomprehensible without recourse to transcendence, making the statement tautological, since the human is defined precisely by the relationship with transcendence. However, by saying that death, thinking, forming political entities or ethics are impossible without recourse to transcendence, are impossible for the animal, are impossible on the plane of (false) immanence, I also delineate key points of interest for a theory of radical immanence. These four areas will offer primary ground for a theory of the immanent animal – seeking aid from different thinkers, I will try to understand if and how one can indeed think, form political entities, act ethically or die within the plane of immanence – in other words, can one think, form political entities, act ethically or die like the (and not necessarily *an*) animal. Much as many of these reasonings will only have a tentative character; hopefully, they will pave the way for the desired way of thinking as well as allow for answering a more basic question – is it even possible?

Before I continue, I need to make two remarks. Firstly, everything I have said earlier about the concept of the animal, the human, as well as thinking, death, ethics etc., will need to be set in a different context. It does not mean that what I said earlier was false – on the contrary, the success of this chapter relies precisely on the validity of the earlier reasonings. However, those familiar concepts I have used in a familiar meaning throughout the text will have to be rethought (or, more consistently speaking, recreated) under different circumstances, on a different plane. In fact – to speak strictly along the lines of Deleuze and Guattari's theory I have used as a methodological framework throughout this text – it is only now that they will truly reveal themselves as actual *concepts*. Of course, keeping the names of the concepts of thought, ethics, politics, death etc., as well as – or maybe especially – those of the human and the animal, is in a way a feint, a tactical move, for the new concepts will be identical with the previous ones in name only. However, in using them, I will try to ask the same questions, or take into account the same events, problems or, for the lack of a better term, areas of experience – in other words, the new concepts share a history with the old ones. Again, the old concepts were not *false* – the truth or falsity of a concept is not relevant – but they all remain linked to a God that limits the possibility of creation.

The question "why should we not limit the possibility of creation of concepts?" is only apparently valid. Of course, it is possible that this freedom of thought will lead to creating monsters. But monsters are only monstrous for someone who has a set view of the norm – a view that is provided precisely by what I identified as God, i.e., by a transcendent reference point. The survival of concepts should not be dictated by any norm outside of the concepts themselves.

Secondly, although this chapter will use inspirations from different philosophers – with a key role reserved to Spinoza – it will only use them, again, tactically. Since neither Spinoza nor Deleuze have proposed a strict link between the concept of the animal and immanence, it is difficult to use the work of any philosopher as a whole, or point to a passage or thought that represents the way of thinking I am trying to propose in a straightforward or wholesome manner. This is not a claim to absolute novelty – indeed, I will be heavily dependent on inspirations from philosophers and scientists alike. One might say that the concepts are there, but not quite unveiled yet, or not presented in the area I am trying to explore. As Hasana Sharp notes in a similar context, with reference to Spinoza:

> [R]eaders are troubled by the fact that despite his denial that man "through rank and dignity is a being entirely different from *things*, such as irrational animals," he asserts that we can do whatever we like with nonhuman things. Although Spinoza does not place any special metaphysical value on humanity, he urges us to prefer ourselves and one another to other natural things. He appears to authorize the exploitation of nonhuman nature, even as he

excoriates those who fantasize a God who might "direct the whole of Nature according to the needs of their blind desire and insatiable greed."[5]

Sharp's argument is that Spinoza's ontology – which I also will be heavily relying upon on the following pages – allows for a non-speciesist approach, or one that allows the treatment of humans and animals on the same level. It will only be visible later that this approach does not necessarily abolish all the things that we thought human – meaning simply that it is not "reductive Cartesianism" like Dawkins's theory I analyzed in the preceding chapter – but allows the categories that are traditionally thought of as human to be rethought as not specific or not essential in the sense that they would form an essence of humanity, and thus be exceptional to man.

However, Sharp goes on, Spinoza does not go on to actually propose such a non-speciesist approach, but stops short of it, saying that indeed we, humans, should refrain from abusing animals, or that at least we should treat (practically and theoretically, if it is indeed a valid distinction) animals and humans on the same level. Asking why he is doing that is not the aim of this study – one might say that he was too deeply rooted in his time (a "socio-historical" explanation) or perhaps wanted to preserve his habits (a psychological one): all this is irrelevant – what is important (also for Sharp) is that he provides an ontological framework for the way of thinking I am trying to develop – in the language of this work I would say that Spinoza's ontology enables the creation of the concept of the animal based on a plane of (radical) immanence, while he himself does not explore this possibility.

Again – this remark (neither mine nor, as I believe, Sharp's) is not intended as one belonging to the history of philosophy, claiming, for example, that Spinoza is the first "true philosopher of immanence," or, on the contrary, that he failed to achieve radical immanence and I intend to go "further than Spinoza." My problem is different than that of Spinoza's, and in his thinking (and that of others) I find a tool to solve it. The issue is not *who Spinoza was*, but *what we can do with his help* – the aim is not to expose his philosophy, but to answer the question "how can one think (act ethically, die, etc.) as the animal?" – one that Spinoza never himself answered or even posed in such a way.

Roughly the same arguments could be made for other thinkers mentioned in this chapter – rather than try to expose any complete, let alone systematic, way of thinking about the animal or animals, I will try to follow some of their thoughts that allow thinking the animal in terms of a strict plane of immanence.

The thinking animal

The analysis of Kant's approach to man as an *animal rationale* allowed the conclusion that thinking – at least the kind of thinking that separates the human from the animal – always, in the end, falls back on God as its necessary condition. This is a special case of a wider and popular philosophical assumption that thinking needs to be somehow linked to truth – it is irrelevant here what kind of definition of truth we are discussing, as no matter which truth it is (even a seemingly "immanent" version saying that coherence within a system is the condition of truth), it still serves as an external referent to thinking itself. Gilles Deleuze cites the affinity of truth and thinking as one of the elements of the classic image of thought:

> The philosopher, it is true, proceeds with greater disinterest: all that he proposes as universally recognised is what is meant by thinking, being and self – in other words, not a particular this or that but the form of representation or recognition in general. This form, nevertheless, has a matter, but a pure matter or element. This element consists only of the supposition that thought is the natural exercise of a faculty, of the presupposition that there is a natural capacity for thought endowed with a talent for truth or an affinity with the true, under the double aspect of a *good will on the part of the thinker* and an *upright nature on the part of thought*. It is because everybody naturally thinks that everybody is supposed to know implicitly what it means to think.[6]

It is precisely this ease and supposed natural character of thinking, of the claim that we need to think "right" and "true" (i.e., that we need to follow rules of this or that "logic" to come up with "true" thoughts) that makes us stop short of saying that the animal (as we perceive it philosophically) could also "think" – and that we can think in this way.

The first condition of thinking as the animal is therefore to rid ourselves of the supposition that thought has a necessary relationship with truth, that it needs to be judged for its truthfulness or falsehood – or, in fact, that it needs to be judged at all, at least if judgment means posing oneself in a transcendent (even relatively transcendent) position. All judgment is, in a sense, the "judgment of God."[7]

If (animal) thinking does not want the truth (what Deleuze called "good will" of thinking), if it should not be judged for its truth, then what can it be? It simply needs to be something that *happens*, and happens as necessarily and as accidentally as any other event, including physical ones. This, of course, is a Spinozist thought, and it is Spinoza who I will be following in trying to delineate what "animal thinking" should be. I will start with a short exposition of what Spinoza says about the necessity of thinking, and only then will move to show how it helps us to "think as the animal."

Spinoza's is a philosophy of absolute necessity[8] – nothing can come to being and no change can occur without a cause, and every cause determines a certain effect. There is nothing accidental or miraculous in Nature, everything is causally connected. This regime is exactly the same for both thoughts and bodies – "The order and connection of ideas is the same as the order and connection of things."[9]

This statement deserves a twofold explanation – one concerning the relationship between thoughts and things (Spinoza's "parallelism") and another, explaining the seemingly simple concept of necessity, which will be crucial for the understanding of the "animal" way of thinking.

According to Spinoza, we perceive two types of things, and two types of things only – thoughts and bodies.[10] Thought and extension are two attributes of God or Nature – it is possible that there are more than the two attributes (in fact their number should be infinite), but it is certain that we are able to perceive only those two. It has been discussed if the difference between these attributes is real or only subjective, but this ontological question is less important than another one – about the relationship between the two. Spinoza makes a crucial point in the second Definition of Part I of the *Ethics*:

> A thing is said to be finite in its own kind when it can be limited by another thing of the same nature. For example, a body is said to be finite because we can always conceive of another body greater than it. So, too, a thought is limited by another thought. But body is not limited by thought, nor thought by body.[11]

This means that the causal order of bodies is independent of the causal order of thoughts – we cannot properly say that a thought *caused* a material thing to do so-and-so, neither can we say that As Hasana Sharp explains:

> The notion of parallelism, attributed to Spinoza by Leibniz, misleadingly suggests that for each material "thing" in the world, there is a representation, or idea, in the universal intellect, or mind of God. For the human being, parallelism implies that there are ideas in the human mind reflecting each body part. The imagery of parallelism, as others have pointed out, is thereby easily misconstrued, since it suggests that ideas exist in order to represent bodies and are valid insofar as they do so accurately. Since the attributes are metaphysically independent of one another, ideas and bodies do not express or explain one another but rather express one and the same order and connection of causes.[12]

In a way, this manner of thinking about the body–mind relationship can be comprehended with the help of a structurally similar idea present in Gilbert Ryle's work. Over three centuries after Spinoza, Ryle still sees the classical understanding of the body–mind relationship as "the dogma of the Ghost in the Machine."[13] Its main points are: (1) everyone has a body that is spatial and a mind that is not; (2) although some say that the mind (soul) can live on after the body is dead, it is certain that the two are somehow "harnessed together."[14] Although it is not Descartes who is to be blamed for the doctrine, it is certainly he who, with his somewhat desperate theory about the pineal gland, did much to start a debate about the nature of this relationship. Many philosophers tried to offer different explanations, some going so far as to negate the very existence of bodies or minds, treating them as mere illusions – but even this is only an extreme conclusion from the classical theory than its abolition.

Ryle proposes something that is far simpler and much more subtle (although at first glance it looks like the subtlety of Alexander the Great's answer to the Gordian Knot problem). He claims that the problem is not the character of the relationship between the body and the mind, but the fact that the basis of this theory – claiming that we can speak of minds and bodies as existing in the same way – is completely wrong. He calls it the "category-mistake."[15] In the philosopher's own words:

> [The mistake] represents the facts of mental life as if they belonged to one logical type or category (or range of types or categories), when they actually belong to another. The dogma is therefore a philosopher's myth.[16]

A category-mistake is basically the inability to properly use concepts in their right contexts. Ryle gives quite a few examples; let me just cite one of them:

> John Doe may be a relative, a friend, an enemy or a stranger to Richard Roe; but he cannot be any of these things to the Average Taxpayer. He knows how to talk sense in certain sorts of discussions about the Average Taxpayer, but he is baffled to say why he could not come across him in the street as he can come across Richard Roe.[17]

It is clear that Doe and Roe are two different people, while the Average Taxpayer is a statistical construct. We cannot meet a statistical construct on the street, the same way as we cannot say (except in metaphor) that we are seeing a melody or hearing green. Or that a body is influencing a mind:

> Minds and bodies belong to different logical categories, they are different manners of speaking about things – we cannot say that there "are minds and bodies", because the phrase "there occur mental processes" does not mean the same sort of thing as "there occur physical processes", and, therefore, that it makes no sense to conjoin or disjoin the two.[18]

This certainly does not mean that either bodies or minds do not exist. Ryle explains that while it is "perfectly proper to say, in one logical tone of voice, that there exist minds and to say, in another logical tone of voice, that there exist bodies,"[19] these two expressions use the word "exist" in two different senses. Similarly, when we say " 'the tide is rising', 'hopes are rising', and 'the average age of death is rising'",[20] we use the word "rising" in a manner that is specific for each of the contexts, and if someone were to say that "three things are now rising, namely the tide, hopes and the average age of death,"[21] they would be obviously using language improperly.

This solution – as well as Spinoza's – has the advantage of at once doing away with the Cartesian version of the mind–body problem, and any misconstrued monism. Much as Ryle insists that concepts related to minds and bodies are of two different kinds, Spinoza claims that minds and bodies are two different attributes of the same substance, i.e., from the epistemological point of view, two different ways of expressing the substance – or of us experiencing it.[22] This is why Ryle is useful in clarifying the argument I am trying to make – the difference between bodies and minds is, for Spinoza, indeed ontologically located on the level of the attributes; however, it needs to be stressed that those attributes are of the same substance – to us, they differ only insofar as we experience them. No process or law which governs the behavior of the modes of one attribute can be absent in the other, and vice versa; there can be no additional laws which govern one of the attributes. The fact that we do not notice this lies in our bad linguistic (or more widely – cognitive) habits, which Ryle so well exposes. The difference between thought and body is, from the point of view of the one substance, that of expression: the substance expresses itself as different attributes. From our point of view (as "men"), it is an epistemological difference: we perceive the one substance in two ways, we have two types of language to name it: one for bodies, and one for minds. The mistake we are making is to presuppose that some of our perceptions about our mind (free will, acting towards an end) are its actual traits; the further mistake is that we tend to attribute those traits to the physical world as well.

Spinoza was, of course, ardently opposed to the belief in what is commonly known as free will, and any thought that would make any being, man or God, exempt from the natural, mechanical order of things and able to turn their actions to a fixed end – or, put it simply, the plane of transcendence. It naturally follows from this that man and his actions and thoughts should be analyzed in the exactly same way as the rest of nature. Spinoza's project is, thus, simple: "I shall consider human actions and appetites just as if it were an investigation into lines, planes, or bodies."[23]

Emotions and affects – and thoughts – belong to the same natural order as planes and bodies; there is no substantial difference between bodies and minds, there is only an attributive one, one that hinders on our perception. How does this way of thinking about thinking save us from the Cartesian dogma, the problem of the Ghost in the Machine, or from thinking in terms of the plane of transcendence? In other words – how does this help us to think as the animal?

Before I answer this question, I need to make two further remarks. Firstly, it could be said once again that all this is obvious – we know that man is a part of Nature and the sciences (even the so-called social sciences) have learned Spinoza's (and Bacon's and Galileo's. . .) lesson, and man is analyzed in the proper way. I believe, however, that what I have shown in the preceding chapters is enough to prove that this is blatantly wrong – even when we use terms like "reason," we usually still remain on the old plane, and we still remain as "human" as we can.

Secondly, it is interesting to note that Spinoza proposes an explanation of how thinking in terms of the plane of transcendence comes into being. The answer lies in imagination. It is seemingly one of the points that somebody who does not believe in a soul that acts independently of the body has the most trouble to understand – how can a body experience something that is not present (for this would be "imagination" presented from the point of view of the attribute of extension). Spinoza offers an explanation in Proposition 17 of the second part of the *Ethics*:

> If the human body is affected in a way [*modo*] that involves the nature of some external body, the human mind will regard that same external body as

actually existing, or as present to itself, until the human body undergoes a further modification which excludes the existence or presence of the said body.[24]

This is much easier to understand with the help an example. Say, when I enter a shop and sense a smell that I associated with bread, I will assume that there is bread in the store – but when I look around, seeing that in fact there is no bread in the store (in Spinoza's language: my body will undergo a modification that excludes the presence of bread), I will stop thinking of it as existing in the store (present to myself). Less obvious thoughts can be triggered by more subtle bodily experiences.

Imagination in itself is not a bad thing – it only becomes so when we take things that we imagine as real. Spinoza says this very clearly:

> I should like you to note that the imaginations of the mind, looked at in themselves, contain no error; i.e., the mind does not err from the fact that it imagines, but only insofar as it is considered to lack the idea which excludes the existence of those things which it imagines to be present to itself.[25]

One might easily explain this in terms of the preceding example on the bodily level – when I come to the store and smell bread, and even though I realize that I do not see it anywhere, I still might salivate: this reaction makes no sense unless my body still experiences the inexistent bread as existing. This type of confusion is the source of many mistakes men make in judging Nature (any judgment of this sort is a mistake in itself): not understanding the nature of beings, we can only imagine them, and mistake the order we have given to things in our imagination with the way things actually are.[26]

This tendency is clearly visible in its connection to the creation of so-called transcendental terms (like Being, Thing and Something, a classical scholastic list):

> The human body, being limited, is capable of forming simultaneously in itself only a certain number of distinct images [. . .] If this number be exceeded, these images begin to be confused, and if the number of distinct images which the body is capable of forming simultaneously in itself be far exceeded, all the images will be utterly confused with one another. [. . .] When the images in the body are utterly confused, the mind will also imagine all the bodies confusedly without any distinction, and will comprehend them, as it were, under one attribute, namely, that of entity, thing, etc. [. . .] It all comes down to this, that these terms signify ideas confused in the highest degree.[27]

So, if any transcendental term is in its nature an imagined one and a confused one, and if the wrongful use of imagination means taking imagined terms for real ones, then what about absolute terms like God (or indeed Spinoza's Nature)?[28] It seems that – at least for Spinoza – any belief in absolute, transcendent points of reference is by its nature a confusion. Thinking in terms of the plane of transcendence is thus an effect of a confused imagination – and so is the belief in any even relatively absolute point of reference.

Now back to thinking as the animal. Spinoza's theory is one of immanence, well-understood parallelism and non-exceptionality of any being. It gives a two-fold blow to any Kanteseque theory of reason and of man as a reasonable animal. First of all, it shows that thinking as if God existed (and only this is properly human thinking) is very confused (and confusing) business, suitable only for those that live locked in their imagination. Secondly, it does not treat thinking as an exceptional activity that is possible only for certain beings – usually only humans, but in some cases also "higher" animals, which, as we have seen, does not alter the picture at all – but as a universal phenomenon, one of the attributes of existence.

This second point is clearly crucial. Spinoza puts it thusly:

> man consists of mind and body, and the human body exists according as we sense it. [. . .] what we have so far demonstrated is of quite general applica-tion, and applies to men no more than to other individuals, which are all animate, albeit in different degrees. For there is necessarily in God an idea of each thing whatever, of which idea God is the cause in the same way as he is the cause of the idea of the human body. And so whatever we have asserted of the idea of the human body must necessarily be asserted of the idea of each thing. Yet we cannot deny, too, that ideas differ among themselves as do their objects, and that one is more excellent and contains more reality than another, just as the object of one idea is more excellent than that of another and contains more reality.[29]

Hasana Sharp, in commenting upon this section, rightfully points out a cer-tain unease with which we approach such a theory – that "ideas and minds, for Spinoza, belong to any and all existent beings, be they rocks, cars, birds, or chew-ing gum."[30] But although many can (and do)[31] find it ridiculous, she claims that "[t]his vision of reality yields promising effects even if one may not readily accept that shoes have minds."[32]

While certainly agreeing with Sharp's point, I believe that it would be prudent to make a small correction to the manner of speaking that she adopts. Terms like "shoes have minds" seem to indicate that a mind or thoughts is something one "has" – we are concrete beings who have been endowed with a mind, and so have shoes. This sounds like primitive spiritualism, while Spinoza's (and Sharp's too, even though not expressed in the most fortunate fashion) intention is exactly the opposite – not to say that every being has been given a soul, but that what we call souls, minds or thoughts are in fact as common as bodies, and, on a certain level, simply are the same.[33] Our bodies are affected by external bodies, and in the same way our thoughts are affected by other thoughts. This is one process, only described through two different attributes.

Thinking, therefore, does not have to be conscious, is never "free" in the sense of not being necessary, and does not have to be linked to a "higher power." It is exactly the same as movement and rest, and behaves like "lines, planes and bodies" in animals as much as humans. Of course, this means rocks and shoes, too – Spinoza's theory allows for a truly universal treatment of thinking. The only reason that we cannot imagine a thinking stone – or, more correctly, cannot consider the stone in terms of its thinking other than as not thinking at all – is that we are so different from it (as we can see by the difference of our bodies).

Thinking happens to animals, and so it does to stones and us, as long as we do not believe that we can anchor our thinking in an imaginary entity such as God – this is the core idea behind thinking "as the animal." But there is more – if thought is parallel to the body, then by changing our bodies, we can change our thoughts. We can – and here we move from treating "the animal" as a philosophical concept to "empirical" animals – become-animal in thought simply by changing the way our body is affected. Consciousness as we know it is not necessary for thought, neither is the unconscious – all is needed is to be, to affect and to be affected.

This does not mean that there are no differences in thinking, and that there cannot be different *modi* of thinking, some of them more powerful or more potent than the others. If it is true that we are living, especially in the 21st century, in a "human" world, then it is at least partly due to the exceptional powers of thinking men have; again, this means that our bodies, too, have exceptional powers. I am not trying to deny that "human" thinking – one happening according to a plane of transcendence – is powerful and useful; however, one of its flaws is that it makes us blind to different kinds of thinking. It stops us from allowing thinking to "happen to us," and privileges one kind over others, and the capabilities of one or several types of beings over the others. It makes the world less just and poorer in thought than it actually is.

If this is so, what can a philosopher do to change this state of affairs? So far, the recreation of the concept of thinking has been more or less an abstract undertaking, forming the conditions of possibility for such thought. In the last part of this subchapter, I will attempt to sketch out a more practical approach, which would allow us to engage in thinking as the animal in a more concrete manner.

A seemingly obvious ally in such a quest would be the neurosciences, working from the assumption that thinking is based on physiological processes. However, the problem with the neurosciences is that for scientists working in a big chunk of the field, thinking happens primarily (or only) in the brain. As Alva Noë notes, "establishment neuroscience is committed to the Cartesian doctrine that there is a thing within us that thinks and feels. Where the neuroscientific establishment breaks with Descartes is in supposing that that thinking thing is the brain."[34] This brain-oriented paradigm has two immediate downsides.

Firstly, it limits the taking into account the experiences of animals whose main thinking power does not come from the brain. In other words, the focus on the brain can be another way of putting the caesura between thinking in the "right" (using the brain) and the "wrong" way. While it would be ridiculous to see neuroscientists delve into thinking shoes – or any other beings that have little to do with neurons – there are obvious blind spots when one focuses on the brain to understand thinking. In this context, Cary Wolfe speaks of the "scandal of the cephalopods,"[35] as squids have demonstrated the most spectacular evidence of intelligence outside the mammal (or, more widely, vertebrate) world. And while

Wolfe himself might be accused of rehashing the man/animal caesura (searching, as he does, for "the capacity to respond"),[36] we can certainly agree that our thinking cannot be guided by "generic biological markers, such as membership in a particular species or phylum."[37]

Secondly, this brain-centered paradigm limits the very notion of the body to the brain itself, which not only greatly impedes the range of any endeavor to engage in thinking as the animal, but also seems problematic when it comes to what we ourselves know as thinking or – in the case I am about to discuss – consciousness. What I am referencing is extended cognition and – in particular – Alva Noë's book *Out of Our Heads*, which makes a compelling argument about how consciousness depends on not only the brain, but also "on my history and my current position in and interaction with the wider world."[38] In Noë's view, rather than being possible brains in vats, we are always already involved with what is outside, and we are essentially open to the outside: "We are patterns of active engagement with fluid boundaries and changing components."[39] In Spinoza's terms, one might say that our consciousness – along with the rest of our thinking – is shaped by our entire body's affects, and not just by those belonging to a specific part of it.

If not by help from neuroscientists, how can a philosopher give justice to this type of thinking? How to actively change one's affects in order to think as the animal? One possible answer lies in reversing a strategy that was used in Chapter 4 to better understand the concept of the animal *via* the concept of the human. There, the *loci* of the human–animal caesura were analyzed to show how the distinction always relies on the concept of the human being created on a plane of transcendence. Now, we need to get back to one of these *loci* and re-interpret it in a way that can serve as a privileged spot on the plane of immanence rather than open us up to an external referent.

In the context of wondering what it means to think – which, as we have already seen, means wondering what it means to have language, and *a fortiori*, what it means to be human – Heidegger points out that "the animal has no hand." This is, of course, linked to his analysis of thinking as a form of handicraft, in fact "handicraft par excellence."[40] Heidegger strongly emphasizes that it is not a simple prehensile organ he is talking about – "Apes, too, have organs that can grasp, but they do not have hands"[41] – and links the hand, for example, to gesture, which is only possible with those who have language. Heidegger's insistence on the essential character of his analysis and his peculiar stress on the hand in the singular – which was profoundly analyzed by Derrida in his *Geschlecht II*[42] – is an instance of a tendency I have already analyzed earlier: namely, to disincarnate the *Dasein* as far as possible. Derrida comments shortly that "hands, that is already or still the organic or technical dissipation"[43] – it is only in the singular that the hand can actually achieve its true purpose, which is to escape the organic (the animal) and to reach out to an external point of reference, thus establishing a plane of transcendence.

However, as Didier Franck remarks when discussing the notions of *zuhanden* and *vorhanden*, "it is essential – although this necessity was something that Heidegger never took into account – that Dasein have hands so that, all metaphors aside, the being of the being that is could be named being-at-hand."[44] In other words, Heidegger's understanding of the hand, regardless of how essential (or disincarnated) he would like it to be, relies on the material reality of the hand (or hands) – and so does thinking. To subvert Heidegger's thinking about thinking, we need to get rid of the notion that a supposedly essential analysis of the hand is possible, and – figuratively speaking – reattach the hand to the body without losing the sight of it being a privileged *locus* for the analysis of thinking. Once the hand stops being instrumental in the establishing of a plane of transcendence, it can again be analyzed simply as a prehensile organ – unique only in its material architecture (and its affective capabilities) and not in any essential way (at least in the classical understanding of the word).

The notion of the hand – or the grasping organ in general – as an opportune area for the analysis of thinking has not been unheard of in biology. Indeed, the cognitive ethologist Frans de Waal suggests that in light of recent discoveries we should wonder if the "manual" kind of intelligence we share with other primates is indeed the *only* kind of intelligence – for example, one of the reasons scientists were so slow at recognizing the full extent of the cognitive powers of elephants was that they failed to construct their experiments for beings who use trunks instead of hands. De Waal goes so far as to suggest that as much as our intelligence and language may have evolved thanks to our manual activities such as throwing stones or spears, the elephants' intelligence might have much more to do with their trunks (also on the neurological level).[45]

While results of inquiries in the natural sciences cannot be of immediate use to philosophy conceived as creating concepts,[46] it is important to see how we – as philosophers – can learn from the shift in thinking that is hinted by de Waal. What we should wonder is not whether if we think – in one way or another – "with our hands" and our hands are unique in the animal kingdom, then we are the only beings who think (it is easy to see that the premises and the conclusion of this reasoning can be easily switched to accommodate other versions of human uniqueness). We should rather ask: if we are used to hand-thinking, then what is it like to trunk-think or tentacle-think (or, perhaps, to branch-and-root-think)? Rather than treat hand-thinking as a model for thinking as such, we should ask in what way our thinking is unique and how we can build upon this uniqueness to discover other ways of thinking – and, conversely, to what extent we are doomed to fail in this respect because of the incongruence of our bodies and other animals' (or plants' or stones') bodies.

The ethical animal

When analyzing Kant's theory of practical reason, I came to a conclusion that ethics, insofar as it separates men from animals, needs to be built upon thinking in terms of a plane of transcendence. The same was true for Levinas, even given the vast difference between the two theories.

In the realm of ethics, thinking in terms of transcendence has an effect of introducing a separation – this is especially important to note given the professed universalism of these two theories, which, in the case of Kant, hinges on the supposed universality of reason, and in the case of Levinas', on the supposed universality of the face (regardless of nationality, creed, gender etc.). And even if it might be argued that in some cases the line is not drawn between all humans on the one hand, and all animals on the other, we are dealing with a radical frontier between the human and the animal.

In the previous subchapter, I have suggested that – when introducing a perspective of radical immanence – we need to consider "animal" thinking as a universal phenomenon, thus including not only the idea of thinking animals, but also stones, shoes etc. This, of course, did not mean that shoes are sentient or that stones have souls in the Cartesian sense of the term. In the same way, ethics established within the concept of the animal would not – at least in a straightforward manner – be an *animal* ethics. We already know that while the concept "animal" has a lot to do with "empirical" animals; one cannot simply be equaled with the other. However, immanent ethics, based upon universality rather than separation, would *a fortiori* have a place for animals.

Here, again, Spinoza's *Ethics* comes as an invaluable source of inspiration, starting from its idiosyncratic title, which lead some thinkers to propose alternatives which supposedly better convey the idea behind Spinoza's philosophical project – or at least point us towards a fruitful interpretation. For Guido Ceronetti, this term was "medicine."[47] For Deleuze, it was "ethology."

The idea behind understanding the *Ethics* as medicine is clearer once one remembers what has been said in the previous subchapter. Traditional ethics is a matter of the soul – one of its most important traits or claims is the struggle between the soul and the body, or the struggle against certain bodily urges – and it is not just present in religious or "platonic" ethics, but also, as we have seen previously, in the case of Kant, who denies any action the character of a truly moral one unless it is not "pathological" in any way. For Spinoza, this view is clearly wrong, as we cannot say that the soul can control the body in any way; they are two sides of the same reality.

In a sense, it means that provided we do not talk about bodies as if they were souls and vice versa (we do not make "category-mistakes"), it does not matter if we speak of one or the other dimension. However, since we have so many more superstitions about souls than we have about bodies, Spinoza proposes to take the body as a model for ethics. In this sense, "Medicine" is thus an ethics of the body.

In a similar vein, Deleuze proposed calling Spinoza's *Ethics* "ethology." As Hasana Sharp notes, Deleuze's understanding of the notion is peculiar, but faithful to Jakob von Uexküll's.[48] Deleuze points to a number of defining traits of ethology that go far beyond the traditional definition as "study of animal behavior." Firstly:

> Ethology is [. . .] the study of the relations of speed and slowness, of the capacities for affecting and being affected that characterize each thing.[49]

Ethology is thus the study of bodies as Spinoza defines them (primarily as capable of movement – hence the "speed and slowness" – and entering relations). Each body has characteristics which allow it to be affected in a certain way (an eye is affected by light but not by sounds) and to affects others in a certain way (a hand can touch but not shout).

Secondly, ethology considers

> the way in which these relations of speed and slowness are realized according to circumstances, and the way in which these capacities for being affected arc filled.[50]

This means that the bodies are considered in terms of how they are affected in given circumstances – are they given more or less power, are they destroyed or made bigger, are they poisoned or rejuvenated?

And lastly,

> ethology studies the compositions of relations or capacities between different things [. . .] it is a question of knowing whether relations (and which ones?) can compound directly to form a new, more "extensive" relation, or whether capacities can compound directly to constitute a more "intense" capacity or power. [. . .] How do individuals enter into composition with one another in order to form a higher individual, ad infinitum? How can a being take another being into its world, but while preserving or respecting the other's own relations and world?[51]

This last point is crucial for the "ethical" value of ethology, which becomes an art or science of knowing which bodies can form bigger communities with other ones, and which have a decomposing effect. To better understand this, we need to understand the way bodies are defined in this approach, as it is one of the things that makes Spinoza's philosophy (at least in its Deleuzian interpretation) so unique. Let me start with another citation:

> animals are defined less by the abstract notions of genus and species than by a capacity for being affected, by the affections of which they are "capable," by the excitations to which they react within the limits of their capability. Consideration of genera and species still implies a "morality," whereas the Ethics is an ethology which, with regard to men and animals, in each case only considers their capacity for being affected.[52]

This could be a definition of other objects as well as animals; however, the notion of ethology is a zoological one, so the description is fitting. The resignation from defining objects in terms of genera and species, and instead proposing to do it in a relational manner, has a triple effect. Firstly, even though an object may be thought to have an essence,[53] it is not an abstract entity that has a questionable relationship with the "actual" object (a Platonic idea), but a list of relations it is capable of. This was covered by the first definition of ethology, and has profound implications for how we see the world. Deleuze's most famous illustration of it is that

> there are greater differences between a plow horse or draft horse and a race-horse than between an ox and a plow horse. This is because the racehorse and the plow horse do not have the same affects nor the same capacity for being affected; the plow horse has affects in common rather with the ox.[54]

It is clear why this is the case – a plow horse and an ox are good for pulling things, while the racehorse is good for running fast. From this, there stems the second characteristic of this way of defining objects – it is always relative, tactical. This might be a circumstantial trait that is founded on our inability to count all the affects of such complicated creatures as horses or oxen, but the fact is that since we cannot do it, we will always have to stick with a limited number of those we choose to take into consideration. As seeing beings (affected by light of a certain wavelength), the ox, the plow horse and the racehorse are roughly the same, but when defined in terms of usefulness (as Deleuze does, following Freud's Little Hans), some of them are more like others. This makes a certain prudence necessary in the exercise of ethics. I will come back to this point.

Thirdly, the understanding of ethics as ethology proposes a "universal" vision of being where no object can be judged "better" or "worse" – in fact, the objects stop being absolute wholes or individuals in the etymological sense of the term. From the point of view of ethology, there is no essential difference between a leg and a human being. A human being and a rock can be defined in the same terms – all one needs to do is to count the affects they are capable of. This is why the Spinozo-Deleuzian ethology is so useful for an ethics of the animal, based on a plane of radical immanence. There is no single trait on which ethics can be based – not for humans, not for humans and some animals, and not for all living beings. No affect – be it the ability to reason, to have religion, or to speak – is in any way "better" than the others. The only border between beings is an "affective incongruence." There could be an ethics for human beings, animals, hands, cars and stones, all equally valid – but certainly human ethics is naturally more human-oriented, and so on.

The notion that beings are not "absolute wholes" also means that, in a sense, they are all parts – and all of the parts have parts. Naturally, this pertains to both bodies and souls:

> The human body is composed of very many individual parts of different natures, each of which is extremely complex.[55]
> The idea which constitutes the formal being of the human mind is not simple, but composed of very many ideas.[56]

A part of a body or soul also has its parts, and bodies and souls are parts of larger bodies or ideas; finally, they are part of everything there is. Therefore, Spinozian ethics is not only universal but also multifocal – a body needs to be considered not only in its relationships to other bodies of the same type (stone-ethics, leg-ethics), but also in the possibilities of connections it might have to other bodies that it is a part of.

This naturally takes us back to the last aspect of ethology discussed previously. Let me cite the fragment again:

> How do individuals enter into composition with one another in order to form a higher individual, *ad infinitum*? How can a being take another being into its world, but while preserving or respecting the other's own relations and world?[57]

This problem can easily be transformed into a moral commandment, which would sound more or less like this: "Act in such a way that when you affect another being, do it only in order to form a higher community, or at least not to destroy their existing relationships to the world." And while one may think this might be a good starting point for a morality of immanence, sadly, it is not – a "morality of immanence" is a *contradictio in adjecto*. The animal is not moral – but it might be ethical.

What we *can* do is try to live our lives without abusing our imagination – without treating its objects as existing, or even thinking *as if* they existed: in a word, without building a plane of transcendence. There is no higher criterion we can establish to guide us in our ethical endeavors, as there are no general laws that are applicable in singular cases. Rather, with a Spinozan or "animal" ethics we will always operate in conditions of uncertainty, because ethics must be based on knowledge, and this knowledge is never full. As Spinoza famously says:

> [n]obody as yet has determined the limits of the body's capabilities: that is, nobody as yet has learned from experience what the body can and cannot do, without being determined by mind, solely from the laws of its nature insofar as it is considered as corporeal. For nobody as yet knows the structure of the body so accurately as to explain all its functions, not to mention that in the animal world we find much that far surpasses human sagacity, and that sleep-walkers do many things in their sleep that they would not dare when awake clear evidence that the body, solely from the laws of its own nature, can do many things at which its mind is amazed.[58]

Deleuze comments:

> The approach is no less valid for us, for human beings, than for animals, because no one knows ahead of time the affects one is capable of; it is a long affair of experimentation, requiring a lasting prudence, a Spinozan wisdom.[59]

Ethics is thus an experimental affair – however, this "experiment" is very far from what the so-called experimental sciences claim it to be: a practice that guarantees repetition, provided that all factors are controlled. It is closer to an artistic experiment, one where a certain process is set in motion and takes its immanent course, and the course is different every time, because we never know the exact factors, the affects that need to be taken into consideration. In this way, it is closer to madness than science, since the same actions can produce different results.

When we leave the imagined plane of transcendence that gave us a false feeling of security in our ethical judgments, the radically immanent animal ceases to be like the falsely immanent animal, if the latter – as Descartes used to say – always acts mechanically and perfectly. Leaving the territory of supposed animal certainty, we stumble upon the territory of experiment and uncertainty, populated by bodies of whose possibilities and potentialities we are ignorant. The radically immanent, ethical animal is first and foremost an uncertain animal.

One might ask about the stakes of such ethics – what does this uncertainty make us capable of? Preaching about an "ethical community to come" aside, the one thing it requires is an openness – and it shows the possible extent of such an openness. We see the world as a set of beings we take for granted – a horse is one being, a tree is another, a stone is yet another. What traditional ethics or morality does is to tell us how we should limit our behavior towards those beings as classes, while not challenging us to look at them individually, or, better said, specifically – there is no such thing as a horse or a stone, there are only multiplicities of affects, and each of them is different, allowing different combinations to arise and requiring a different approach to be elevated in its affective possibilities. The corollary to saying that we do not know what a body can do is that it can always do more than we, in the way we see the world – limited by, at least in this example, our ethical scope – can imagine it doing. In a way this prudence tells us: forget all you know about the class, the traits, the uses of the being you are approaching and let it affect you – and see what happens. In the words of Ronald Bogue,

> The duty to the other (if one must speak of duty) is to affect and be affected, to suspend, as much as one can, the categorization and comprehension of the other, and then open oneself to the undetermined, hidden possible worlds that are expressed in the affective signs of the other.[60]

The reluctance to speak of duty or moral commandments might be misunderstood as a refusal to actively engage in practical matters, thus putting those who speak of (radically) immanent or animal ethics in the awkward position of preaching the art of experiment and shying away from any action. While this objection is in itself misconstrued, it does show the fine line one needs to walk between prescriptions for concrete experiments and introducing a transcendent referent – this is perhaps best described in the opposition between "recipe" and "theory" proposed by Philippe Pignarre and Isabelle Stengers. While a (scientific) theory is concerned with generalization and "explaining why [things] 'work' in terms that transcend the situation in which they 'work',"[61] a recipe involves producing the conditions of possibility for events that empower those who engage with these recipes, even though they cannot ascertain that a given recipe will work every time. And while for Pignarre and Stengers the term "empowerment" is used clearly in a "humanist" political context, it seems fitting – also because of the Deleuzian inspirations which are dear to Stengers – to use it in a more Spinozo-Deleuzian vein, which would mean translating it into making oneself capable of more affects.

As in the previous subchapter, the recipe I would like to propose is based on a reinterpretation of a *locus* that was for the thinking of transcendent ethics in terms of immanence. Previously, it was Heidegger's concept of the hand; this time, I shall focus on Levinas' notion of the face. Here, as previously, some interpretative abuse is needed to make Levinas' theory into a recipe – most importantly, the absolute or paradigmatic character of the face-to-face relation has to be done away with.

We have already seen how this paradigmatic character operates within Levinas' theory. The epiphany of the face produces a transcendence which forces me to act in a certain way – minimally, forbidding me from killing the Other. However, this epiphany is only possible when I encounter a human face; should an ethical relationship be possible with regards to other beings, it is only on the condition that it is modeled on the human face paradigm – I have already cited the fragment where Levinas points this out: "The human face is completely different and only afterwards do we discover the face of an animal. [. . .] the prototype of this is human ethics."[62] Any attempts to reinterpret the French phenomenologist's theory so that it can incorporate animals have usually included the belief that the ethical is first and foremost a human matter based on the epiphany of the face. Such is the case, for example, in Matthew Calarco's *Zoographies*. On the one hand, Calarco opposes the somewhat romantic notion that animals may also have access to transcendence, but on the other, he seems to hold on to the idea that animals and humans can, in principle, find themselves on the same ethical field, which is defined from the point of view of the human ethical experience. In his own words:

> The human, then, is an ethical concept rather than a species concept; conse-
> quently, the concept of the human could – at least in principle – be extended
> well beyond human beings to include other kinds of beings who call my ego-
> ism into question.[63]

Although his analyses push him towards a conclusion not entirely foreign to the ideas presented in this subchapter, underscoring the element of uncertainty, prudence – or indeed risk – involved in ethics, as well as their universality and the inclusion of not only animals and humans, but also perhaps "dirt, hair, fingernails and ecosystems,"[64] Calarco stops at the crucial point of expanding the notion of the ethical so that it is not limited to human ethics or face-ethics.

I believe that it is through a thorough, *ethological* reading of Levinas that providing the conditions of possibility for such an expansion is possible, and such a reading needs to treat the face-to-face relation as a recipe rather than a paradigm. As with the reading of Heidegger in the previous subchapter – and the readings in

the chapter on "Anthropology" – this manner of interpreting does not mean con-
tradicting Levinas, but showing the localized nature of his analyses. It is certainly
true that the face of the Other affects me; it may even be true that its affecting me
can first and foremost be described by the commandment "thou shall not kill."
However, in an ethological reading, this commandment does not describe a universal moral obligation, but only a localized way I am affected by the face of *an* other.

Rather than provide moral guidelines, ethics as ethology seeks to understand
the specific ways of how bodies can affect each other and how they react to being
thusly affected. This way of understanding ethics can be followed with practical
guidelines which, as with the previously mentioned question of thinking, can be
formulated with the help of a reading of biological (or ethological, in the contemporary biological sense of the term) texts. By these I do not mean subscribing to
the deeply humanistic efforts of such ethologists as Frans de Waal, who show that
morality in a sense close to what we see in humans also exists in primates[65] – these
developments are undoubtedly scientifically interesting and politically useful, but
do little to change our understanding of what ethics can be. Rather, ethics as ethology would need to engage in a detailed analysis of the actual ways different beings
can be affected.

To provide but one starting point for such an undertaking – and a particularly
Levinasian one at that – let me point to Elizabeth A. Tibbetts and her research on
wasps. Tibbetts found that certain species of wasps – mostly those who live in
groups – are surprisingly good at recognizing faces, a skill that was once attributed only to higher mammals. More interestingly, while the ability is usually used
for recognizing wasp "faces" (the unique black patterns they have on the front
of their heads), the insects can be trained to recognize human faces as well.[66] It
is perhaps enlightening to note how Tibbetts and her coresearchers measure recognition – it is by putting two wasps together in a container and analyzing their
aggression towards one another.[67]

What this means is that a wasp is a particularly "Levinasian" creature, changing
its behavior based on being affected by another's "face." While for a scientist the
main questions that arise from such research may focus on the neural correlates
for such facial recognition or the usefulness of these findings for the development of facial recognition software,[68] for an ethologically minded philosopher,
the Levinasian wasps point to a number of ethical questions, such as "What is it
like to be affected by a wasp-face?", "How can I change my body to be affected
by more than just human faces?" or – to paraphrase Ronald Bogue – "How can
I open myself to the 'hidden possible worlds' of wasps and other creatures"?

This type of questioning not only opens us up to new affects, but also puts into
question the implicitly hierarchical view of the animal kingdom and our place in
it – while the previously cited efforts of de Waal and others are based on a global
evolutionary similarity which holds only for the closest "relatives" of man, engaging with the ethics of wasps and other animals usually referred to as lower on the
evolutionary scale (such as certain anemones)[69] can rearrange our whole understanding of what it means to be "similar" to other beings, leading to new possible
classifications and affective alliances.

The political animal

As we have seen with Heidegger's example of bees, a closer scrutiny of what the definition of man as "political animal" entails in the case of Freud, or Schmitt's analysis of the concept of the political, the difference between man and animal is based on a reference forming a plane of transcendence. It seems to make little difference if the problem is analyzed on the ontological (Heidegger), psychological (Freud) or strictly political (Schmitt) level; we tend to come across the same structure.

It is puzzling in its own way why psychoanalysis – the school of thought responsible for our being used to thinking man as internally split, as containing a difference – would be so reluctant to find its own patients dreaming of being multiple, dreaming of being not just one, but many (wolves, for example).[70] This is a truly humanist trait that lies under the mask of an appreciation of human–animal likeness or substitution. In fact, no such thing is at stake here – as Deleuze and Guattari showed, the psychoanalytical animal is symbolical and "humanized" precisely by its individuation:

> [I]ndividuated animals, family pets, sentimental, Oedipal animals each with its own petty history, "my" cat, "my" dog. These animals invite us to regress, draw us into a narcissistic contemplation, and they are the only kind of animal psychoanalysis understands, the better to discover a daddy, a mommy, a little brother behind them (when psychoanalysis talks about animals, animals learn to laugh).[71]

These animals are only here for the Oedipus complex they satisfy or symbolize, are used in a purely tactical manner, and while this itself is no accusation (I am doing precisely the same thing in this work, albeit searching for an entirely different effect), the tactic here is a humanist one – rather than shake the all-too-human vision of human–animal relationships (also on a conceptual level), it makes the animal humanism's useful idiot.

But how is this all linked to politics? A "human" political community has a twofold characteristic. First of all, as we have seen, it is based on a reference to a sovereignty or an enemy, a reference that forms a plane of transcendence. Secondly, and this is the more important part, the political community is formed by humans, that is, subjects that have been cut out from reality in a certain sense. Although men's hands, their shoes, hats and belly buttons are "inside" the political community, they do not "form" it, and are incapable of the needed reference. More importantly, the animal in man is incapable of forming such a bond. This is why any multiplication or internal division of man is unthinkable for Freud, as it is for any humanist. Only man as an individual can form a truly political community. Any internal multiplication or division is a rupture.

An entirely different picture of political communities can be presented with a recourse to Spinoza. Following a hint from Antonio Negri, who claims that "the political thought of Spinoza is to be found in his ontology, meaning in the *Ethics*, much more than in any other parallel or posterior work,"[72] I will not focus on works such as the *Political Treatise* or the *Tractatus Theologico-Politicus*, but rather try to understand the rudiments of the metapolitical vision present in the Dutch philosopher's opus magnum.[73]

While we are certainly used to seeing ourselves as individual beings who form communities, I have already shown in the previous subchapter that it is not necessarily so for Spinoza. The universal character of the projected animal ethics stemmed directly from the assumption that a man's (and any other) body is composed of a large number of parts and, conversely, it is a part of other different structures. There is more – one can think reality in a wholly different manner, by composing structures or creatures that are wholly different from how we usually see it. In this way, the concept "man" is only an effect of one way of categorizing; so is "political community" – and the very logic behind this type of categorizing is built upon the plane of transcendence. When we try to think upon a plane of immanence, new categorizations, concepts, and communities emerge.

It is obvious that this view is a challenge to the way we see the world and applying it seems difficult, especially because we are used to seeing individuals where we could see multiplicities and seeing divisions where we could see communality – and surely we would become less self-assured and more prone to act like a dog who chases its own tail: but in fact, who are we to say that the tail is its own if the dog itself does not seem to think so?

If there is one tendency in contemporary post-humanist thinking that can be understood in Spinozian terms, it is the movement's interest in the notion of symbiosis; however, as I will try to show shortly, using Spinoza's ontological framework in this context radically changes the way we need to look at these problems.

The preoccupation with symbiosis seems to stem from two sources. The first, more general one, is the idea – usually associated with the work of Bruno Latour – that rather than dividing the world into the realms of things and people, we should treat human and non-human actors equally and analyze the networks they co-produce. The political stakes of this theory seem significant to Latour himself, who, near the end of *We Have Never Been Modern*, writes about the "Parliament of things" and compares his endeavor to that of his "predecessors [. . .] [who] invented rights to give to citizens or the integration of workers into the fabric of our societies."[74] In other words, on a political level, we need to recognize the importance of the beings previously reduced to the status of "things," be it animals, plants or nonliving objects, for the shape of our community.

The second inspiration behind the preoccupation with symbiosis is the discovery of the symbiotic origin of multicellular life (as well as certain organelles inside the cells themselves), leading to a complication of the sociobiological belief in competition as the only driving power behind evolutionary processes – as Brian Massumi puts it, "[t]he law of competition has had to bow before a healthy dose of cooperation."[75] The key scientists responsible for these developments are Lynn Margulis, whose work focused mainly on the role of symbiosis in the development of plant and animal cells, as well as James Lovelock, who, with his contentious Gaia hypothesis, helped to re-imagine Earth as a self-regulating, symbiotic organism.[76]

Aside from the aforementioned Brian Massumi, thinkers who are partial to the importance of symbiosis to their ontological or political vision include Donna Haraway, Rosi Braidotti and Alphonso Lingis; regardless of the many obvious differences – style notwithstanding – between the three, there is a common shape of the argument that can be reconstructed from their works.

The arguments usually start from an example of symbiosis in nonhuman animals – Alphonso Lingis' paragraph about Brazil nuts and bees is a prime case in point:

> The flowers of Brazil nut trees can be pollinated by only one species of bee. This bee also requires, for its larvae, the pollen of one species of orchid, an orchid that does not grow on Brazil nut trees. [. . .] The Brazil nut tree is hardwood, and the husk about its seeds is of wood hard as iron. There is only one beast in Amazonia that has the teeth, and the will, to bore into that husk. It is a medium-sized rodent [. . .] Without that rodent, the nuts would be permanently entombed, and Brazil nut trees would have died out long ago.[77]

In Haraway's narrative, the example of the dog–woman relationship in the context of companion species plays the same role.

The next, crucial step is to present this kind of relationship as at universal in nature ("There is perhaps no species of life that does not live in symbiosis with other species")[78] and – *a fortiori* – encompassing also men. As Haraway puts it,

> "Companion species" is a bigger and more heterogeneous category than companion animal, and not just because one must include such organic beings as rice, bees, tulips, and intestinal flora, all of whom make life for humans what it is – and vice versa.[79]

Those symbiotic categories extend to more than just biology, as they encompass also historical and linguistic relations, among others, thus transcending the usually well-guarded boundaries between nature and culture (thus existing in the realm of nature-cultures).

This approach – sometimes with references to a Whiteheadian ontology – puts relations before subjects, claiming that it is only thanks to the former that the latter are created, thus putting into question the Cartesian lone ego cogito, a foundational myth of modernity.

The perspective I am seeking to employ here – the creation of the concept of the animal on a plane of immanence – is not far away from what was described previously. Indeed, at least Braidotti quite explicitly reveals the close affinity of her thinking to Spinoza, whom she calls her "favourite philosopher";[80] Deleuze and Guattari are also not far from the interest of the three aforementioned thinkers of symbiosis, regardless of the many points on which they disagree. However, a Spinozian approach as described in this chapter differs in a few crucial points from the one previously described – I believe that, at the least, it complements the symbiotic argument, and perhaps even offers a new interpretation of what symbiosis means altogether (however counterintuitive it might seem).

When we start from the key notion that each body at once is made up of different parts and makes up other bodies (it is, at least for the sake of this discussion, irrelevant how or where this process ends, on either of its sides), then the problem of relation itself needs to be redefined. For the symbiosis theorists, the question is, primarily, "How is it that two (or more) subjects shape each other in their relation?"; for example: "How is it that a woman and a dog emerge in the course of the companion species relationship?" (in the case of Haraway) or "How is it possible that the Brazil nut tree, the bee and the rodent have emerged in the course of the symbiotic relationship that allows all three to survive?" (in the case of Lingis). A more Spinozian way to pose this question would be different. Since all the bodies in question already are parts of the same substance, we do not need to ask about the origin of their relationship, but rather about the fact that we see them as separate subjects in the first place. Why do we tend to see the dog and the woman, or the wasp and the orchid as separate beings and need significant epistemological effort to appreciate their codependence, rather than see the dog–woman or the wasp–orchid aggregate as entities in the first place?

While a more thorough response needs to wait until the examination of the Spinozian take on the problem of essence, conducted in the next subchapter, all the work in this book up until now has led to suggesting at least a partial answer in the case of the human and the animal. These effects – the separation of man/woman and dog, of man/woman and his/her intestinal flora, or of man and rice and bees – can perhaps only be obtained with a strong selective mechanism that forms a framework enabling such differentiations. I have been trying to show – with the help of Deleuze and Guattari in the chapter "Polemology," as well as Agamben in "Anatomy" – that the plane of transcendence, on which the concept of the human is created, is one such mechanism. If so, it is because of this mechanism that we not only recognize the human in such a way as we do – not just separate from the outside of man, but also strategically differentiated from the immanent "animal within" – but also that we only see humans as capable of forming political communities. In a way, this produces a vicious circle – only the human is capable of producing political communities, but the communities are political only if they are produced by humans.

From this view, several consequences stem for "animal politics" or politics of radical immanence, making it parallel to animal ethics. The first, perhaps most obvious one is the postulate of universality. On the one hand, there is no ultimate way to *a priori* determine the content of such a community, which may contain humans, animals, as well as intestinal flora, bees, Brazil nuts or any other bodies. On the other hand – and more significantly – there is no way to *a priori* determine the character of the subjects of such a community – they may be what we call humans and animals, but also human–animal assemblages or nut-bee–rodent relationships.

The second consequence is that, as with the case of ethics, animal politics always remain an experimental affair – the political communities that emerge can only be described rather than assessed. With the description being based on the affects that a community is capable of, the difference between communities and subjects fades and shows how dim the difference between ethics and politics was in the first place. Such a description must take into account that there are no fixed essences of object, and the same object can have different traits – or indeed *be* different – in various political circumstances. Moreover, there are no possible political commandments; as Massumi puts it, "[a]nimal politics recognizes no categorical imperative. It lives the imperatives of the given situation, immanent to that situation."[81]

The third consequence of the new interpretation of symbiosis – in itself a consequence of the previous one – is that the only way a community can be accessed or formed is through an experiment or "recipe" as described in the previous subchapter. This allows for reinterpreting the previously described theorists of symbiosis – especially Haraway and Lingis – as recipe-providers or experimenters rather than ontological thinkers. While, ontologically speaking, the relationship between the woman and the dog is always present, epistemologically speaking, we must attune ourselves in a way to be affected by the other, whether it be a dog, rice or a bee, thus multiplying things which our body can do. There is certainly no denying that everything is connected, but this remains a platitude – even if, following Spinoza, we call this interconnected whole "substance" – without a concrete affective engagement or openness that will actually allow us to appreciate the political value of this interconnectedness. It is only through proposing recipes for such engagement that a Spinozian can escape the trap of, as Braidotti describes it, "the holistic fusion that Hegel accused Spinoza of."[82]

Rather than focus on the recipes already found in the works of post-humanist thinkers mentioned in this chapter, I would like to suggest exploring one that has – to my knowledge – not yet been analyzed in the context of animal politics or animal studies in general, namely Jean-Luc Nancy's idea of the "inoperative community."

Nancy's thought about the community – aside from a slight humanistic bias which I will point to towards the end of this subchapter – is well aligned with the objective of this study for several reasons, the most important of which is the radical rejection of any "plane of transcendence" (to use Deleuze and Guattari's language) in the creation of a true community.

The negative point of reference of Nancy's idea is the presupposition that a community has a "substance" or a "goal" that is supposedly internal to it, but in fact forms a transcendent point of reference.[83] This mythical "foundation" of community (one thinks of the Freudian myth of the murder of the father of the horde) or "sovereignty" that hovers above the community like the ghost of a long-dead king (or a theological presence, as Schmitt teaches us) forms precisely that reference, and is always presupposed as such. Therefore, this community (although Nancy does not seem to be interested in such a conclusion) needs to be formed out of beings who are capable (either essentially or cognitively) of appreciating such a presupposition. It is therefore a typical "human" community, a "political" one in the sense of man as a political animal.

True community, on the contrary, has no presuppositions; it is created in a whole different logic. The original title of Nancy's book, *La communauté desœuvrée*, points to the word *œuvre*, meaning a work of art, but one that was planned from the beginning, on a blueprint or in the mind of God or man, and created along with this plan. The work of art as *œuvre* is the exact opposite of what I identified as an "artistic experiment" in the preceding subchapter. One might say in a similar vein that the *communauté desœuvrée* is an experimental community.

This is one of the meanings of Nancy's notion that the community is not *pre-supposed*, but *exposed* – its character cannot be known in advance, it is only knowable insofar as it shows – exposes – itself. But an inoperative community is also a community of beings who expose themselves to others. And in the first place those beings expose themselves in their finitude. As Nancy says:

> A singular being *appears*, as finitude itself: at the end (or at the beginning) with the contact of the skin (or the heart) of another singular being, at the confines of the *same* singularity that is, as such, always *other*, always shared, always exposed [. . .] Community means, consequently, that there is no singular being without another singular being, and that there is, therefore, what might be called, in a rather inappropriate idiom, an originary or ontological "sociality" that in its principle extends far beyond the simple theme of man as a social being (the *zoon politikon* is secondary to this community). For, on the one hand, it is not obvious that the community of singularities is limited to "man" and excludes, for example, the "animal."[84]

It seems, thus, that for Nancy, community of the exposed singular beings is an originary experience which on the basic-most level precedes the singularity of those beings, regardless if they are referred to as "men," "animals" or others. Only in the community, allowing the exposition to the other, does a being realize its finitude. And since exposition is mutual, the beings always appear to each other, they always already communicate. As Nancy says:

> [finitude] appears, it presents itself, it exposes itself, and thus it *exists* as communication. In order to designate this singular mode of appearing, this specific phenomenality, which is no doubt more originary than any other [. . .] we would need to be able to say that finitude *co-appears* or *compears* and can only *compear*.[85]

In another place, Nancy says that compearing is based on articulation – one being appears to another insofar as it articulates. This articulation is not necessarily linguistic – it is a "sharing of voices"[86] that do not need to be *understood* or share a consensus, but are to signal the singularities of the beings involved – though Nancy does not say that explicitly, it seems that the articulations form the equivalent of animal cries which signal position or mark territory. To put it in the terms used earlier in this subchapter, it seems that for Nancy, compearing is based on affect in the basic sense of the term – we only compear so long as we are able to affect and be affected by the other, and it is in this affecting and being affected that we co-create each other as members of the community, regardless of our being men, animals, or belonging to other groups.[87]

That alluding to animals is not entirely out of place here is testified in certain points of Nancy's own text. I have already cited one such place, but the next citation is crucial for the matter:

> the community of articulation cannot be simply *human*. This is so for an extremely simple but decisive reason: in the true movement of community, in the inflection [. . .] that articulates it, what is at stake is never humanity, but always *the end of humanity*. The end of humanity does not mean its goal or its culmination. It means something quite different, namely, the limit that man alone can reach, and in reaching it, where he can stop being simply human, all to human.
>
> He is not transfigured into a god, nor into an animal. He is not transfigured at all. He remains man, stripped of nature, stripped of immanence as well as of transcendence. But in remaining man – at this limit (is man anything but a limit?) – he does not bring forth a human essence. On the contrary, he lets appear an extremity upon which no human essence can take place. This is the limit that man is: his exposure – to his death, to others, to his being-in common. Which is to say, always, in the end, to his singularity: his singular exposure to his singularity.[88]

Here Nancy himself seems to be at the limit. On the one hand, the quoted fragment reveals an urge to break up with traditional humanism – there is no human essence or nature, no pre-established characteristic or trait of man. We have already seen how Nancy objects to the notion of man as a political animal; as Fred Dallmayr shows, Nancy can also be read in opposition to another important definition of man, namely *homo faber*, who creates community by creating his own (human) essence through work.[89] But still, this strange being, "stripped of nature, of stripped of immanence as well as of transcendence" somehow remains "man," although man on the limit, on the windowsill of his own . . . essence that is not essence? Nature that is not nature?

What it seems to boil down to is the question of finitude. On the one hand (and this, as I will try to show in the next subchapter, is a key matter), finitude is not limited to a temporal characteristic – it is not essentially linked to death, but also to multiple other, commonplace liminal situations – being-with-others, singularity itself. Also, it is not just a characteristic of man: animals and gods also arrive at this limit, and are equally threatened in their identity upon arriving there.

However, Nancy traces finitude back to Heidegger and seems to fall all too easily into the trap of traditional categorizations (animal/man/god, identity as animal/man/god). This lingering attachment is what stops us midway – and so does the very title for the community-forming exercise, "literary communism." Much as it is supposed to work like a provocation, it still is a negative exercise; even though "literature" is not understood here in the traditional sense, only human men would be provoked, or at least "provokable," by this enunciation. In other words, it is only a certain type of human who has the capability of affecting and being affected by anything that is "literary" in any sense of the term.[90]

Beyond Nancy's humanistic bias however, there lies a very "animal" form of community – one that is not formed by an exclusion based upon a supposed essence, but upon non-articulated articulations of appearing to one another – affecting and being affected by one another – and upon a finitude that does not provide a quasi-essence for one kind of beings, but is a shared predicament that allows communication and communality in the first place. It is a matter of bodies exposed to one another in a shared space.[91] But it is the aspect of this shared finitude that is the most important one of Nancy's contributions to this work. The next and last part of this chapter will endeavor to explain why.

The dying animal

As we have seen in the chapter concerning his approach to the human–animal caesura, Heidegger's claim that animals do not die, but simply perish, is far from stating that there is something unique about the event of human physical death which makes it more special than the death of an animal. Dying is rather a way of being man, and only in being-towards-death is *Dasein* truly itself. I have tried to show that this way of thinking and the way it differentiates man from other beings stems from obstinately creating the concept of the human upon a plane of transcendence with death as an absolute reference point (even if the referent itself cannot necessarily be said to exist).

Heidegger's philosophy forms a negative framework for the goal of this subchapter – an "animal" thinking of death would have to reject the kind of exceptionalism of death he proposes.[92] This needs to be done on two grounds – firstly, death needs to be thought of as universal, at least insofar as of its many forms none makes the being who suffered it essentially different from another. Secondly, death should not be thought of as different from other forms of finitude. I have already touched upon this topic with Nancy, who showed that death is *one of* the limits (along with, for example, being with others or singularity) of not only man, but also the animal or god. On the next few pages I will try to elaborate on these two points.

An obvious inspiration for any inquiry about the universal nature of death is biology, or rather theoretical (and often popular) texts inspired by biological accounts. A good example of such a book is Bernd Heinrich's *Life Everlasting: The Animal Way of Death*,[93] where the author shows how in the animal world, death – although a final event for the being which succumbs it – is always an opportunity for (certain forms of) life to flourish.[94] While this type of narrative does away with the exceptionality of death by putting it in its rightful place in the ongoing biological processes, it seems to also erase the subjective component of dying, imposing a circle-of-life narrative which vacillates between detachment and mystique.

A similar biological inspiration fuels Braidotti's thought; however, the author of *The Posthuman* clearly sees the problematic repercussions of an all-too-positive approach to death (and life). Braidotti finds her solution in a reading of Deleuze, who

> suggests that to make sense of death, we need an unconventional approach that rests on a preliminary and fundamental distinction between personal and

impersonal death. The former is linked to the suppression of the individual-ized ego. The latter is beyond the ego: a death that is always ahead of me and marks the extreme threshold of my powers to become. [. . .] Because humans are mortal, death, or the transience of life, is written at our core: it is the event that structures our time-lines and frames our timezones, not as a limit, but as a porous threshold. In so far as it is ever-present in our psychic and somatic landscapes, [. . .] death as a constitutive event is behind us; it has already taken place as a virtual potential that constructs everything we are.[95]

I believe this fragment is a good summary of where Braidotti's interests lie in her approach to the problem of death. I will try to analyze its several key points. Firstly, the distinction between personal and impersonal death mirrors Agamben's distinction between *bios* and *zoe*. In her discussion of the concept of life, Braid-otti seeks to empower *zoe* over *bios*, effectively engaging with a more inclusive concept of life than the human(ist) *bios* – in the language of *The Open*, it might be said that Braidotti, appreciating the split nature of man, situates herself on the side of the animal rather than on the side of the human. Indeed, Braidotti herself seems to encourage such an interpretation when she says that life "is impersonal and inhuman in the monstrous, animal sense of radical alterity: *zoe* in its powers."[96] However, by seeing life as the radically other, thus, somewhat paradoxically, in a negative way (at least epistemologically), she ventures into the realm of the mystical (even going so far as to call life "cosmic energy").[97] More importantly, it seems that by virtue of this radical inflation, the concept of *zoe* loses its sharpness and focus. The following excerpt is a good example of this tendency:

"Life" in the sense of "*bios/zoe*" is a fundamentally amoral force, the true nature of which is best expressed in its relentless generative power. Cells reproduce and carry on, no matter what. There is no implicit a priori differ-ence between cancer and birth, or between a malignant proliferation of cells in cancer and the benign probation induced by pregnancy.[98]

The problem is that not only life is a/has "relentless generative power." Biologi-cally speaking, viruses are not alive – and nonetheless they "reproduce and carry on, no matter what." Mountains get larger, tides and shorelines are locked in a fight over who gets to grow, and the Universe would not exist in the way it does were it not for the "relentless generative power" of stars, thanks to which heavier elements were formed from hydrogen and helium. Braidotti would perhaps agree that all the aforementioned things are "alive," given her preoccupation with a planetary perspective, earth's agency and becoming-earth; at the very least, she indicates that "a focus on the vital and self-organizing powers of Life/*zoe* undoes any clear-cut distinctions between living and dying."[99] The Spinozian inspiration here is clearly visible – Braidotti's world is driven by a somewhat more creative version of *conatus* – but it seems that in an ontology in which virtually everything is alive, nothing actually is. I will come back to this problem at the end of this chapter.

The second important point of Braidotti's approach to the subject is that – contrary to the tendency I criticized while analyzing Heinrich – she does not forget the importance of the subjective dimension of death and intensely criticizes the "inhuman" practices of today's necropolitical regimes; in short, as she points out, she "does not deny the reality of horrors."[100] Conversely, she does not idealize life or subscribe to an uncritical "pro-life" agenda (either of the conservative or liberal kind), arguing "against the Christian-based belief in the alleged self-evidence and 'worth of life'."[101] For all her leanings toward the mystique, Braidotti is adamant and clear-headed on this point, and staying as clear from the residue of the religious belief in the sanctity of "all" life, as she does from a more secular "idolatry of the natural order."[102]

The final point I would like to touch upon here is the notion of death as an "ever-present," "constitutive" event that is "written in our core." Granted, while the understanding of life as *zoe* means that death is but a "porous threshold," taking into account the "reality of horrors" means that a proper place needs to be found for it, even ontologically. Agreeing on only an ego-centrical notion of death would fall into the Heideggerian trap of construing a plane of transcendence with death as an absolute referent, and therefore "sacraliz[e] death as the defining feature of human consciousness."[103] This does not mean that its presence cannot be simply erased. For Braidotti, therefore, the very fact that we are aware of our mortality means that death has always already happened – "[d]eath is the event that has always already taken place at the level of consciousness."[104] In this way, firstly, the life/death distinction is blurred on an epistemological level, and secondly, the mortal human can be understood as desiring death in the desiring of a full life (rather than one that would simply last). Braidotti strikes a particularly Freudian note when she says that "[w]hile at the conscious level all of us struggle for survival, at some deeper level of our unconscious structures all we long for is to lie silently and let time wash over us in the stillness of non-life."[105] It seems, therefore, that while Braidotti negates the quasi-Heideggerian teleological notion of death as something that pulls us on the conscious level,[106] it does push us towards life; on the unconscious level, however, we are still defined or at least motivated by the desire to die on our own terms, to seek out our proper death as we do with our life; this means reinstating death as a transcendental point of reference (perhaps even doubly) only this time, the plane of transcendence is created on the level of the unconscious. Moreover, the notion of death as ever-present in our consciousness as a past event seems psychologically implausible – I will come back to this point with the analysis of Spinoza.[107]

It appears that Braidotti's philosophy of death is caught in a double bind between two conflicting – and readily important – tendencies: the first is to universalize and naturalize death, while the other is not to lose sight of it as a psychological and ontologically important event. The danger of the first is the understanding of life as a mystical cycle of equally unimportant changes (what one might call a "*Lion King* ontology"); the danger of the second is the absolutization of death *à la* Heidegger. While Braidotti seems to be acutely aware of these two dangers, she appears not able to fully escape the both of them. Rather than treat it as a contingent problem (e.g., with the logic of the argument) I would like to treat it as an essential one – it might indeed be impossible to profess a vitalist ontology and treat death seriously, or to treat death seriously and not think it absolutely.

Any "animal" thinking of death – thinking death on the plane of immanence – has to take this double bind into account. In an endeavor to satisfy this claim, I will start with a reading of Deleuze and especially Spinoza, which strays from the one proposed by Braidotti in important points.

Deleuze had a problematic approach to death, as he seemed to put it out of the perspective of his thinking – or rather, composed his thinking in a way that whenever it would have to think death, it would seemingly reach a problematic limit, cease to be the thinking of pure immanence. Let me cite two crucial fragments. First one is from his famous last text, *Immanence: A Life*:

> We will say of pure immanence that it is *a life*, and nothing more. It is not immanent to life, but the immanence that is in nothing is itself a life. A life is the immanence of immanence: absolute immanence: it is complete power, complete beatitude.[108]

The second is from the book *Spinoza: Practical Philosophy*:

> All that we call bad is strictly necessary, and yet comes from the outside: the necessity of accidents. Death is all the more necessary because it always comes from without. [. . .] It is death's necessity that makes us believe that it is internal to ourselves. But in fact the destructions and decompositions do not concern either our relations in themselves or our essence. They only concern our extensive parts which belong to us for the time being, and then are determined to enter into other relations than our own. [. . .] in reality, what is involved is always a group of parts that are determined to enter into other relations and consequently behave like foreign bodies inside us.[109]

It is the apparent incompatibility of these two fragments that lies at the heart of the problem, and helps pursue a more promising version of thinking death on the plane of immanence. The incompatibility in question stems from the fact, that, on the one hand, on the plane of immanence, there is no "outside" (neither is there an "inside"); on the other hand, if immanence is life, then (as the fragment from the Spinoza book tells us) death can only come from the outside. There is certainly no place for death on the plane of immanence, but neither is there a place for it outside it, as either there is no such place, or it is not a plane of immanence.

It is easy to understand why Deleuze would be so keen on keeping death external, given his strong opposition towards any ontology that would make of death or the death drive a sort of an internal transcendent, which would in any way determine (or indeed produce) human essence. I have already indicated why the positioning of a transcendent referent "inside" is a wrong way for producing a plane of immanence. But it seems that neither does placing death so radically "outside." This is not to say that from any point of view, death is not real – the decomposition of beings does happen, and there is only a certain amount of change that what we call a being of a certain kind can bear to still be called a certain being – this point was already emphasized during the analysis of Braidotti's approach. Death is also external, although in a way that does not violate the rule of immanence.

Before I go further, I need to return to Deleuze's source text, namely Spinoza's *Ethics*. For Spinoza, the terms "good" and "bad" are always relative, and there is no objective "good" or "bad" in Nature.[110] If he keeps those terms, it is only for tactical reasons – to use them in relation to a projected "human nature." What he is going to be using these words *for* is assessing changes in the human being, changes of character that do not lead him to cease being a human being (that is, non-essential changes), but only those which make him more "powerful."

This reservation – that the changes are not essential – means that essential changes are possible, otherwise Spinoza could not give us his disturbing, if hyperbolic, example ("a horse is as completely destroyed if it changes into a man as it would be if it were to change into an insect").[111] How is this to be understood?

To establish that, we need to comprehend what Spinoza means by "essence." In the *Ethics*, he uses a quite classical definition, according to which essence is the aspect or thing by which a thing is what it is, or – and this will be the more important – is *conceived as* this certain thing.[112] This means that essence is both an ontological and an epistemological category.

From the ontological point of view, essence can be equated with the *conatus*, a force "with which each thing endeavors to persist in its own being is nothing but the actual essence of the thing itself."[113] Of course, this definition should not be read "psychologically," at least not in the sense that a being consciously strives to stay what it is; it means that a stone stays a stone as long as it can, and a horse stays a horse.

From the epistemological point of view, the problem is more complicated. Spinoza tackles the issue of how an essence is formed thusly:

> [S]o many images are formed in the human body simultaneously (e.g., of man) that our capacity to imagine them is surpassed, not indeed completely, but to the extent that the mind is unable to imagine the unimportant differences of individuals (such as the complexion and stature of each, and their exact number) and imagines distinctly only their common characteristic insofar as the body is affected by them. For it was by this that the body was affected most repeatedly, by each single individual. The mind expresses this by the word "man" and predicates this word of an infinite number of individuals. [. . .]
>
> But it should be noted that not all men form these notions in the same way; in the case of each person the notions vary according as that thing varies whereby the body has more frequently been affected, and which the mind more readily imagines or calls to mind. For example, those who have more often regarded with admiration the stature of men will understand by the word "man" an animal of upright stature, while those who are wont to regard

a different aspect will form a different common image of man, such as that man is a laughing animal, a featherless biped, or a rational animal.[114]

In other words, the way we form universal terms is based on our experience and it is a singular matter – this is why a horse is a horse (for us) only insofar as we perceive it to be one, and, conversely, we have learned to call horses because the entity-horse has striven to keep itself together (*conatus*) for a time that is long enough for our bodies to gather experiences that would allow for the creation of this term (be affected by horses long enough).

If we remember what has been said here about Spinoza and ethology, things become interestingly complicated. One of the cornerstones of ethics understood as ethology was that we can never know what a body is capable of. So, since we can only understand essence insofar as we understand affects, we will never be able to fully determine an essence of a thing, and, conversely, we will never be quite sure of when it loses this essence and becomes something else entirely – or perishes.

This interpretation of Spinoza gets us quite close to von Uexküll, showing Deleuze was right in reading the former with the help of the latter. Von Uexküll seems to be pointing to a similar meaning of essence, when discussing the example of the oak:

> In the hundred different environments of its inhabitants, the oak plays an ever-changing role as object, sometimes with some parts, sometimes with others. The same parts are alternately large and small. Its wood is both hard and soft; it serves for attack and for defense. If one wanted to summarize all the different characteristics shown by the oak as an object, this would only give rise to chaos. Yet these are only parts of a subject that is solidly put together in itself, which carries and shelters all environments – one which is never known by all the subjects of these environments and never knowable for them.[115]

While it is not clear whether Deleuze himself would agree with the latter part of this quote – that a tree (or indeed any other subject) remains "solidly put together in itself," seemingly meaning that it does, indeed, have an essence in the classical sense of the term, it is certain that von Uexküll points here to the fact that we – or any other subject – will never know exactly what an object is (capable of), and any consolidated view of that object remains impossible or straightforwardly chaotic.

This is true not only for trees, but for more "paradigmatic" subjects as well: I cannot tell what Paul's essence is if I cannot know all his affects, and even then, Paul may affect me in a certain way that is only reserved for me: for example, if Paul is my colleague, he will affect me principally by his profound interpretations of Freudian psychoanalysis, but his boxing sparring partner might consider his right hook (which we both chose he would not affect me with) as essential. Now this complicates the idea behind the workhorse/racehorse comparison introduced earlier in this chapter – since Paul for me is a scholar, and for his partner a boxer (among other things), for me, he will have more in common with other scholars, and for him, with other boxers: he will belong to another category of beings altogether. What is more, if Paul loses the fragment of his long-term memory which is responsible for his knowledge of Freud, he will undergo an essential change to me – but, as long as he does not lose his right hook, he is still Paul for his partner. Conversely, should he lose his right hand, he will be Paul to me, but not to his partner. To me, Paul will be equally destroyed when he loses his memory as if he were to turn into a horse.

One might think that there is one being who has a privileged access to all of Paul's affects, namely Paul himself. Yet, this is not that simple, even though Spinoza does suggest that "nothing can happen in that body without its being perceived by the mind."[116] This might lead one to think that all affects of the body are consciously known by the soul – however, such a statement would be blatantly false from the point of view of everyday experience. In fact, Spinoza means something much simpler – it is just his theory of parallelism applied to the human body and soul: for every affect of our body, there is a parallel affect of our soul, just as every material stone is also a "mental stone" whose history is exactly the same, albeit in a different attribute. This does not mean that we have an adequate, conscious "knowledge" of our body. In fact, it is impossible, as "The human mind does not involve an adequate knowledge of the component parts of the human body."[117] This negative proposition has its positive counterpart in the preceding one: "The mind does not know itself except insofar as it perceives ideas of affections of the body."[118]

In other words, we only know ourselves insofar as we know what our body can do – and we can never know that adequately. Our body strives to keep itself together, and we make sense of what this "together" stands for only insofar as we perceive its affects. And, of course, this gets even more complicated as we see ourselves "as"[119] something or someone. To Paul, delivering a right hook might be more important than studying Freud. He would be destroyed without the first. Also, what he might not know, he would be destroyed by many other things.

Understanding that, we can venture a reinterpretation of the fragment which inspired the quotation from Deleuze's book a few pages ago: "No thing can be destroyed except by an external cause."[120]

The proof of this proposition goes as follows:

> the definition of anything affirms, and does not negate, the thing's essence: that is, it posits, and does not annul, the thing's essence. So as long as we are attending only to the thing itself, and not to external causes, we can find nothing in it which can destroy it.[121]

Therefore, we can see that the externality of death or destruction is always only relative to the thing in question – when Paul is defined by his ability to interpret Freud, it is clearly not this ability that will stop him from being Paul. If we perceive Paul as a man, we cannot think of his "death-as-man" as anything other than inhuman. However, when we start perceiving Paul – like the exercise that this chapter is thought us to do – as multiple, as "animal," we will have to understand that his death as the death of one or more of the things he is may well lie in another of the things that he is. His multiplied body may well be capable of a "death affect": one of the parts that is in him might destroy others that we deem essential for Paul; as Deleuze noted, "what is involved is always a group of parts that are determined to enter into other relations and consequently behave like foreign bodies inside us."[122]

This way of thinking death makes it always relative or relational, but allows a way to think it as a non-essential affect. Neither of these characteristics make death any less painful or frightening – in fact, we will strongly fight any external affect that endangers what we perceive as our essence – and understanding the relativity of death does not make one a believer in any mystical unity with Nature or Life. It is rather a modest attempt to put death in its rightful place from an ontological point of view: as an event like many others, a destruction of an essence. The shattering of a world might be tragic, but it is hardly a metaphysical scandal.

This interpretation also has profound implications for the concept of life and defining what we mean by living (in more than one sense of the term). Let me address two of these implications. Firstly, I would like to get back to a point I touched upon when analyzing Braidotti's theory of life: namely, the question of the psychological plausibility of the idea that our consciousness should be ever-marred by the past event of our own death; this implausibility – which, in itself, can always be treated as an idiosyncrasy – can be ontologically reinforced by a Spinozian argument. For the author of *Ethics*, keeping oneself alive might be a condition of possibility of many other things in life – including any virtue, which makes it a primary virtue – but it is rarely an occupation in itself. In fact, the thought of death rarely comes into the mind of a healthy, wise person. Spinoza illustrates this with the example of eating:

> The sick man eats what he dislikes through fear of death. The healthy man takes pleasure in his food and thus enjoys a better life than if he were to fear death and directly seek to avoid it.[123]

This is an individual example of a general rule: "Through the desire that arises from reason we pursue good directly and shun evil indirectly."[124] While this might sound like a psychological statement, it is in fact more than that. If we understand death as the greatest evil, we will see that what actually happens is that a wise man does not usually (or ideally) refer to death in what defines his constitution and actions – only somebody driven by fear or anxiety does that. As Spinoza says, "a free man, that is, he who lives solely according to the dictates of reason [. . .] thinks of death least of all things, and his wisdom is a meditation upon life."[125]

It is important to remember that Spinoza's "free man" has no connection whatsoever with what might be identified as a Kantian free man, and the "dictates of reason" are not the dictates of Kantian transcendental reason with its necessary transcendent referent, but rather leading his life in accordance to the rules of Nature.[126]

Spinozian living in harmony with Nature is much akin to what Montaigne understood by the same concept – or indeed to Nietzschean *amor fati*. We have already seen that this attitude is not what we defined as "human" in the earlier chapters – it is in fact the hallmark of the animal way of being. Man was defined by an essential relationship with death, only *was* insofar as he *was dying*, only *was* insofar as he related to the external and absolute possibility of death. The animal, with the simplicity of an analytic *a priori* proposition, only *is* insofar as it *is*. If it is true that the animal is not capable of dying, it is because it does not form a relationship with death as a transcendent point of reference – neither on an ontological nor on a psychological level.

The second implication of a Spinozian theory of death allows to understand the indefinite article used in the title of Deleuze's *Immanence: A Life*, and the citation I started this analysis with. The indefinite article indicates that each life is a singularity rather than an individuality, and something simple and plain rather than an ontological monstrosity of "vitalism."[127] Contrary to a biologistic interpretation, what everybody (including, in a way, himself) sees in the dying Dickensian "scoundrel" is not an instinct or a drive to life, but rather his "singular essence,"[128] stripped from any psychological traits or social status. The dying man is for the first time seen as a being unique in his singularity, something that was until that time clouded behind his repugnant actions; contrary to an Agambenian intuition, he does not become "*homo sacer*," but rather elicits an ethical response. If, as Braidotti would like it, vitalism is the reinforcement of *zoe*, we might say that in this fragment Deleuze is as far from vitalism as ever.

However, it seems that from a Spinozian point of view I have been trying to employ here, a point of view of radical immanence – of the animal – ending with simply equating "singular essence" with "a life" is insufficient, for both the notions of life and death. I have stressed it multiple times before that death is only one form of finitude, no more special than any other. This, of course, does not mean that it is no different – nor that there are not many different deaths: my own death is different from the death of a loved one, of a distant acquaintance, of a pet or a farm animal. Again, this view of death does not try to negate our perception of death – our own or somebody else's – as something less dramatic, tragic or evil. What it does try to do is to set it in a perspective where death neither defines us as human, as it is common to all living beings, nor drives our actions in a necessary manner – in both cases it is only secondary. What is more, it puts it, from an ontological point of view, on the same level as other essential changes, forcing us to focus more on understanding the singular affective make-ups of different beings than on imagining we can posit an individual point of reference that everyone is subjugated to in the same way.

Conversely, saying that death is "not special" also leads to wondering if life, in itself, is so special – either as something that opposing death or as a force of growth and the multiplication of power. Anything that *is* strives to keep its essence – a stone as well as an animal – the *conatus* does not discriminate; everything that *is* grows in the right environment. While we call one thing living, and another not living (dead or otherwise), these categories or concepts are as produced or constructed as any others. We find them useful in the same way that we find classifying one type of beings as thinking, and others as not much so – but, taken as ontological categories, they might be less useful, and certainly less magical, than we often think. Rather than focus on such distinctions, we should focus on thorough analysis of affects that different beings are capable of. In a word, it seems that the animal, which "does not die," does not care if it lives, either.

Notes

1 The term "radical immanence" is also used, among others, by Rosi Braidotti [e.g., *The Posthuman* (Cambridge: Polity Press, 2013), 56; Aristides Baltas, *Peeling Potatoes or Grinding Lenses: Spinoza and Young Wittgenstein Converse on Immanence and Its Logic* (Pittsburgh: University of Pittsburgh Press, 2012), 1]. While both of these authors employ this term in a Spinozian sense, I believe that a recourse to Hagglund enables me to provide a clear definition of how the concept should be understood in the framework of this chapter.
2 Martin Hagglund, *Radical Atheism: Derrida and the Time of Life* (Stanford: Stanford University Press, 2008), 1.
3 This notion is discussed at length for example in: Jacques Derrida, *Aporias: Dying – Awaiting (one another at) the Limits of Truth*, trans. Tomas Dutoit (Stanford: Stanford University Press, 1993).
4 Hagglund, *Radical Atheism*, 1.
5 Hasana Sharp, *Spinoza and the Politics of Renaturalization* (Chicago: University of Chicago Press, 2011), 186.
6 Gilles Deleuze, *Difference and Repetition*, trans. Paul Patton (New York: Columbia University Press, 1994), 131.
7 Cf., e.g., Gilles Deleuze, "To Have Done with Judgment," in *Essays Critical and Clinical*, trans. Daniel W. Smith and Michael A. Greco (London: Verso, 1998), 126–35.
8 "From a given determinate cause there necessarily follows an effect; on the other hand, if there be no determinate cause, it is impossible that an effect should follow." Baruch Spinoza, "Ethics," in *Complete Works*, trans. Samuel Shirley (Indianapolis: Hackett, 2002), 218 (I, A5).
9 Spinoza, "Ethics," 247.
10 See Spinoza, "Ethics," II, A5, 244: "We do not feel or perceive any individual things except bodies and modes of thinking."
11 Spinoza, "Ethics," 217.
12 Sharp, *Spinoza and the Politics of Renaturalization*, 28.
13 Gilbert Ryle, *The Concept of Mind* (London: Routledge, 2009), 5.
14 Ryle, *The Concept of Mind*, 1.
15 Ryle, *The Concept of Mind*, 6.
16 Ryle, *The Concept of Mind*, 6.
17 Ryle, *The Concept of Mind*, 7.
18 Ryle, *The Concept of Mind*, 12.
19 Ryle, *The Concept of Mind*, 12.
20 Ryle, *The Concept of Mind*, 12.
21 Ryle, *The Concept of Mind*, 12.
22 It is crucial to understand that, as Pierre Macherey writes: "The attributes are [. . .] identical to the substance, and likewise the substance is the same thing as its attributes: it is only from the point of view of the intellect that a distinction between substance and attribute can be established, which means that this distinction has no real nature but is only a distinction of reason." [Pierre Macherey, "The Problem of the Attributes,"

in *The New Spinoza*, ed. Warren Montag and Ted Stolze (Minneapolis: University of Minnesota Press, 1997), 71]

23 Spinoza, "Ethics," 278.

24 Spinoza, "Ethics," 256.

25 Spinoza, "Ethics," 257.

26 "And since those who do not understand the nature of things, but only imagine things, make no affirmative judgments about things themselves and mistake their imagination for intellect, they are firmly convinced that there is order in things, ignorant as they are of things and of their own nature." Spinoza, "Ethics," 242.

27 Spinoza, "Ethics," 266.

28 If this is taken seriously, it shows why this interpretation of Spinoza meets the reservation made by Yovel about how immanence should not be treated as "absolute" (cf. the closing parts of the preceding chapter).

29 Spinoza, "Ethics," 251–52.

30 Sharp, *Spinoza and the Politics of Renaturalization*, 63.

31 Sharp points to Margaret Dauler Wilson's book, *Ideas and Mechanism: Essays on Early-Modern Philosophy* (Princeton: Princeton University Press, 1999), who writes, for example, that (p. 139) "[Spinoza's theory] has extremely exotic implications of its own [. . .] fails to capture ordinary notions of the mental, and that it suffers from internal imprecision and inconsistency. In short, it does not work at all."

32 Sharp, *Spinoza and the Politics of Renaturalization*, 64.

33 Perhaps it needs to be underlined that the universality of thought in the world has nothing to do with any notion of the existence of *reason* in the world (in a Hegelian sense or other) – if anything, this subchapter seeks to put thought as far from reason as possible.

34 Alva Noë, *Out of Our Heads*, Kindle ed. (New York: Hill and Wang, 2010).

35 Cary Wolfe, *Before the Law: Humans and Other Animals in a Biopolitical Frame* (Chicago and London: University of Chicago Press, 2013), 72.

36 Wolfe, *Before the Law*, 72.

37 Wolfe, *Before the Law*, 72.

38 Noë, *Out of Our Heads*.

39 Noë, *Out of Our Heads*.

40 Martin Heidegger, *What Is Called Thinking*, trans. Fred D. Wieck and J. Glenn Gray (London, New York and Evanston: Harper and Row, 1968), 23.

41 Heidegger, *What Is Called Thinking*, 16.

42 E.g., "the man of the typewriter and of technics in general uses two hands. But the man that speaks and the man that writes with the hand, as one says; isn't he the monster with a single hand?" Jacques Derrida, "*Geschlecht* II. Heidegger's Hand," in *Deconstruction and Philosophy: The Texts of Jacques Derrida*, ed. John Sallis (Chicago and London: University of Chicago Press, 1987), 182.

43 Derrida, "*Geschlecht* II. Heidegger's Hand," 182.

44 D. Franck, "Being and the Living," in *Heidegger Reexamined, Vol. 1. Dasein, Authenticity and Death*, ed. Hubert Dreyfuss and Mark Rathall (New York: Routledge, 2002), 144.

45 See Frans de Waal, *Are We Smart Enough to Know How Smart Animals Are?* (New York: W. W. Norton & Company, 2016), esp. chapter 5, *The Measure of All Things*.

46 See Chapter 2, "Polemology," for a more thorough analysis of the question of the relationship between the sciences and philosophy.

47 Cf. Guido Ceronetti, *The Silence of the Body: Materials for the Study of Medicine*, trans. Michael Moore (New York: Farrar, Strauss and Giroux, 1993).

48 Sharp, *Spinoza and the Politics of Renaturalization*, 211.

49 Gilles Deleuze, *Spinoza: Practical Philosophy*, trans. Robert Hurley (San Francisco: City Light Books, 1988), 125.

50 Deleuze, *Spinoza: Practical Philosophy*, 125–26.

51 Deleuze, *Spinoza: Practical Philosophy*, 126.

52 Deleuze, *Spinoza: Practical Philosophy*, 27.
53 For an interpretation of Spinoza's notion of essence, see the subchapter "The Dying Animal" in this chapter.
54 Deleuze, *Spinoza: Practical Philosophy*, 124.
55 Spinoza, "Ethics," 255.
56 Spinoza, "Ethics," 255.
57 Deleuze, *Spinoza: Practical Philosophy*, 216.
58 Spinoza, "Ethics," 280.
59 Deleuze, *Spinoza: Practical Philosophy*, 125.
60 Ronald Bogue, "Immanent Ethics," in *Deleuze's Way: Essays in Transverse Ethics and Aesthetics* (Aldershot: Ashgate, 2007), 13.
61 Philippe Pignarre and Isabelle Stengers, *Capitalist Sorcery. Breaking the Spell* (London: Palgrave Macmillan, 2011), 133.
62 Tamra Wright, Peter Hughes, and Alison Ainley, "The Paradox of Morality: An Interview with Emmanuel Levinas," in *The Provocation of Levinas. Rethinking the Other*, ed. Robert Bernasconi and David Wood (London and New York: Routledge, 1988), 172.
63 Matthew Calarco, *Zoographies: The Question of the Animal from Heidegger to Derrida* (New York: Columbia University Press, 2008), 65.
64 Calarco, *Zoographies*, 71.
65 See, e.g., Frans de Waal, *Primates and Philosophers: How Morality Evolved* (Princeton: Princeton University Press, 2016).
66 A brief history of Tibbetts' research on the subject is given in: Elizabeth A. Tibbetts and Adrian G. Dyer, "Good with Faces," *Scientific American* 309, no. 6 (December 2013): 62–67.
67 See, e.g., Allison Injaian and Elizabeth A. Tibbetts, "Cognition Across Castes: Individual Recognition in Worker *Polistes Fuscatus* Wasps," *Animal Behaviour* 87 (January 2014): 91–96. While in this paper the researchers are not absolutely sure that the recognition between wasps happens thanks to the recognition of their "faces" and not other traits, other research on wasps (see, e.g., previous footnote) strongly suggests it is the case.
68 Tibbetts and Dyer, "Good with Faces."
69 While not so cognitively endowed as to recognize "faces," certain anemones regulate their aggression as a function of relatedness during contest behavior; interestingly, they seem to most ferociously attack the individuals that are their closest relatives. See Nicola L. Foster and Mark Briffa, "Familial Strife on the Seashore: Aggression Increases with Relatedness in the Sea Anemone *Actinia equina*," *Behavioural Processes* 103 (March 2014): 243–45.
70 A clear example of this critical line can be found in Gilles Deleuze and Felix Guattari, *A Thousand Plateaus*, trans. Brian Massumi (Minneapolis: University of Minnesota Press, 2005), 28, where the French thinkers accuse Freud of ignoring the fact that in his famous dream, the Wolf Man sees a number of wolves (five or six, depending on the version of the story), and never just one wolf, while "every child knows" that these animals usually form packs.
71 Deleuze and Guattari, *A Thousand Plateaus*, 240.
72 Antonio Negri, *Spinoza for Our Time: Politics and Postmodernity*, trans. William McCuaig (New York: Columbia University Press, 2013), 9.
73 This decision is primarily motivated by the fact that Spinoza's political treatises do not problematize the notion of the political as deeply as his ontology does, since they focus on analyzing the notions of sovereignty and different types of governments. Examples of works concerning Spinoza's *Political Treatise* and his *Theologico-Political Treatise* include Antonio Negri's, *The Savage Anomaly*, trans. Michael Hardt (Minneapolis: University of Minnesota Press, 1991) and his collection *Subversive*

Spinoza. (un)Contemporary Variations, ed. Timothy Murphy (Manchester: University of Manchester Press, 2004); Alexandre Matheron's, "The Theoretical Function of Democracy in Spinoza and Hobbes," in *The New Spinoza,* 49–64 or Warren Montag's, "Interjecting Empty Spaces: Imagination and Interpretation in Spinoza's *Tractatus Theologico-Politicus,*" in *Spinoza Now,* ed. Dimitris Vardoulakis (Minneapolis: University of Minnesota Press, 2011), 161–78.

74 Bruno Latour, *We Have Never Been Modern* (Cambridge, MA: Harvard University Press, 1993), 145.

75 Brian Massumi, *What Animals Teach Us About Politics* (Durham and London: Duke University Press, 2014), 1.

76 See Lynn Margulis, *Symbiotic Planet* (New York: Basic Books, 1999); esp. chapter 8 for the links between Margulis' ideas and the Gaia hypothesis.

77 Alphonso Lingis, "Animal Body, Inhuman Face," in *Zoontologies: The Question of the Animal,* ed. Cary Wolfe (Minneapolis and London: University of Minnesota Press, 2003), 165–66.

78 Lingis, "Animal Body, Inhuman Face," 166.

79 Donna Haraway, "The Companion Species Manifesto: Dogs, People and Significant Otherness," in *Manifestly Haraway* (Minneapolis and London: University of Minnesota Press, 2016), 106.

80 Braidotti, *The Posthuman,* 55.

81 Massumi, *What Animals Teach Us About Politics,* 38–39. While there are numerous similarities between my endeavor and Massumi's – including the understanding of animal politics as an experiment or play (p. 89) – there are also significant differences. Aside from the general methodological approach – a Deleuzian/constructivist one in my case and a new-materialist/posthumanist in Massumi's – one of such differences is the approach to the question of affect, equally important in both projects. While Massumi does cite Deleuze's definition of Spinoza's affect (p. 58), he himself seems to be using the term in the sense of "emotion" rather than a general capability to be affected. This, paired with his partiality to the term "instinct," which, however singularly interpreted, points to his subscribing to inversion of the dualistic scheme of animal/human, thus straying dangerously close to the tendencies I described in the chapter "False Immanence."

82 Braidotti, *The Posthuman,* 92.

83 "If I had to attempt to state the principle guiding the analyses in these texts, I might do so by saying this: community does not consist in the transcendence (nor in the transcendental) of a being supposedly immanent to community. It consists on the contrary in the immanence of a 'transcendence' – that of finite existence as such, which is to say, of its 'exposition.' Exposition, precisely, is not a 'being' that one can 'sup-pose' (like a sub-stance) to be in community. Community is presuppositionless: this is why it is haunted by such ambiguous ideas as foundation and sovereignty, which are at once ideas of what would be completely suppositionless and ideas of what would always be presupposed. But community cannot be presupposed. It is only exposed. This is undoubtedly not easy to think." Jean-Luc Nancy, *The Inoperative Community,* trans. Peter Connor, et al. (Minneapolis: University of Minnesota Press, 1991), xxxix.

84 Nancy, *Inoperative Community,* 27–28.

85 Nancy, *Inoperative Community,* 28.

86 Nancy, *Inoperative Community,* 77.

87 It also should be said that co-appearance or compearing is different from recognition in the Hegelian sense of the term. In an interesting paper [Stuart Dalton, "Nancy and Kant on Inoperative Communities," *Critical Horizons* 1, no. 1 (February 2000): 29–50], Stuart Dalton tries to tackle this subject. He explains his objective thusly: "Given that an inoperative community is the consequence not of a deliberate work intent upon realising a unified essence, but instead results from a certain resistance to totalisation on

the part of singular subjects – who are not drawn together and given a unifying concept by their resistance – how is it possible for an inoperative community to become aware of its own existence? In other words, is it possible for a community that recognises the unheard experience of community (which Nancy articulates) to recognise itself as a community, or must it necessarily remain oblivious to its own reality?" (p. 30). Dalton tries to answer this central question by pointing to the Kantian notions of the *sensus communis* and the "purposiveness without a purpose" from the third Critique. It is interesting to see how this question shows that the case of recognition (always invoking a recourse to transcendence) is almost impossibly difficult to answer in the case of Nancy's essay.

88 Nancy, *Inoperative Community*, 77.

89 See Fred Dallmayr, "An 'Inoperative' Global Community? Reflections on Nancy," in *On Jean-Luc Nancy: The Sense of Philosophy*, ed. Darren Sheppard, Simon Sparks, and Colin Thomas (London and New York: Routledge, 1997), 171.

90 In a later intervention, Nancy himself rejects the notion of "literary communism"; however, he seems to do it for reasons different than "literary" being too "human" a term to underscore the universal nature of the community: "I recently used that expression [literary communism]; its equivocal character makes me reject it now. I am not speaking here of a community of letters." [Jean-Luc Nancy, "Of Being In-common," in *Community at Loose Ends*, ed. Miami Theory Collective (Minneapolis: University of Minnesota Press, 1991), 10].

91 The importance of the body to Nancy's idea of community has been widely discussed, notably in Anja Streiter, "The Community According to Jean-Luc Nancy and Claire Denis," *Film-Philosophy* 12, no. 1 (2008): 49–62.

92 For a different interpretation, which, in the case of death, among others, brings Spinoza (who I am again heavy relying upon in this subchapter) and Heidegger closer together (only to steer them apart again in the end), see Antonio Negri, "Power and Ontology Between Heidegger and Spinoza," in *Spinoza Now*, 307–19.

93 Bernd Heinrich, *Life Everlasting: The Animal Way of Death* (Boston: Houghton Mifflin Harcourt, 2012).

94 The tendency is, of course, not limited to the animal world, with dead trees and vegetation being a crucial resource for the sustenance of many species of fungi, plants and invertebrates (especially beetles), not to mention birds such as woodpeckers feeding on the latter. For popular works concerning the subject see, e.g., Peter Wohlleben, *The Hidden Life of Trees: What They Feel, How They Communicate – Discoveries from a Secret World*, trans. Jane Billinghurst (Vancouver: Greystone Books, 2016); Andrzej Bobiec, ed., *The Afterlife of a Tree* (Warszawa and Hajnówka: WWF Poland, 2005).

95 Braidotti, *The Posthuman*, 131–32.

96 Rosi Braidotti, *Transpositions: On Nomadic Ethics* (Cambridge: Polity Press, 2006), 216.

97 Braidotti, *Transpositions*, 216.

98 Braidotti, *Transpositions*, 223.

99 Braidotti, *The Posthuman*, 115.

100 Braidotti, *The Posthuman*, 132.

101 Braidotti, *Transpositions*, 232.

102 Braidotti, *The Posthuman*, 85.

103 Braidotti, *The Posthuman*, 132. It needs to be taken into account that for Braidotti the transition from *bios* to *zoe* seems to go through the first kind of death, the death of the ego, towards an impersonal becoming.

104 Braidotti, *The Posthuman*, 135.

105 Braidotti, *The Posthuman*, 135.

106 While I believe this is an accurate summary of Braidotti's thoughts on the matter, from a Heideggerian perspective death in the sense of being-towards-death is an affair of ontology, not of psychology, so using the terms "conscious" and "unconscious" might be considered problematic.

107 This kind of Spinozian argument could also be directed against Hagglund's reading of Derrida. The author of *Radical Atheism* claims that mortality is "integral to whatever one desires" (p. 111) and insists especially that one cannot desire immortal life, since the desire of life is the desire of something inherently mortal, thus is implicitly the desire of death.

108 Gilles Deleuze, "Immanence: A Life," in *Two Regimes of Madness*, trans. Ames Hodges and Mike Taormina (New York: Semiotext(e), 2006), 385–86.

109 Deleuze, *Spinoza. Practical Philosophy*, 41–42.

110 "There is in Nature neither Good nor Evil, there is no moral opposition, but there is an ethical difference." [Gilles Deleuze, *Expressionism in Philosophy: Spinoza*, trans. Martin Joughin (New York: Zone, 1990), 261].

111 Spinoza, "Ethics," 322.

112 "[T]hat necessarily constitutes the essence of a thing which, when posited, posits the thing, and by the annulling of which the thing is annulled; i.e., that without which the thing can neither be nor be conceived, and vice versa, that which can neither be nor be conceived without the thing." Spinoza, "Ethics," 250.

113 Spinoza, "Ethics," 283.

114 Spinoza, "Ethics," 266.

115 Jakob von Uexküll, *A Foray into the Worlds of Animals and Men with a Theory of Meaning*, trans. Joseph D. O'Neil (Minneapolis: University of Minnesota Press, 2010), 132.

116 Spinoza, "Ethics," 251.

117 Spinoza, "Ethics," 260.

118 Spinoza, "Ethics," 251.

119 It needs to be noted that this "as" is completely different from Heidegger's "as-structure" I discussed in Chapter 4, "Anthropology."

120 Spinoza, "Ethics," 282.

121 Spinoza, "Ethics," 282.

122 Deleuze, *Spinoza: Practical Philosophy*, 42.

123 Spinoza, "Ethics," 353.

124 Spinoza, "Ethics," 353.

125 Spinoza, "Ethics," 355.

126 "For insofar as we understand, we can desire nothing but that which must be, nor, in an absolute sense, can we find contentment in anything but truth. And so insofar as we rightly understand these matters, the endeavor of the better part of us is in harmony with the order of the whole of Nature." Spinoza, "Ethics," 362.

127 Although working in a slightly different framework, focussing on Deleuze's (and Guattari's) Freudian inspirations (death drive) and the displacement of the concept of life thanks to the idea of desiring-machines, Claire Colebrook also sees the indefinite pronoun as something that differentiates Deleuze's brand of vitalism from its more straightforward, biologically oriented counterparts, be it historical or contemporary (Braidotti's, perhaps, included): "Deleuze's passive vitalism is distinguished by this reference to 'a' life, not a life in general that underlies actual beings and that then allows for an ethics or politics of life. It needs to be distinguished from early anti-Cartesian vitalisms that posit a single driving force, precisely because it is without drive; it is actualized in encounters and events – this or that being or character whose existence is contingent. It also needs to be distinguished from what I refer to as the contemporary turn to life, or the contemporary vitalism of living systems: the idea that we can only understand ethics, aesthetics or technological evolution by referring back to organisms and their management of their milieu." [Claire Colebrook, *Deleuze and the Meaning of Life* (London and New York: Continuum, 2010), 136].

128 Gilles Deleuze, *Two Regimes of Madness*, trans. Ames Hodges and Mike Taormina (New York: Semiotext(e), 2006), 387.

7 Uses of immanence

Conclusions and remarks on practice

I started this book by saying that it is not about animals – and indeed "the animal" took us far beyond animals. This is not a surprising conclusion – a recurring theme in the last chapter was universality, the claim that man should not think himself exceptional because of his supposedly human traits. Indeed, they should be thought of as common to all beings. However, even if this communality was achieved by the recreation of the concept of the animal, its reach was much further – we have seen (or created) thinking stones, ethical communities including hands, etc. How did it come to be? How did rethinking the concept of the animal lead not only to the negation of the radical frontier between the animal and the human, but also the postulate of a theory of universal thinking, ethics, politics, etc.?

As I have stressed, the classical concept of the animal has little to do with animals and a lot to do (as I have shown in the first chapter) with man. Following this concept to establish that it was based on the plane of transcendence, and then trying to see what it would look like when formed based on the rules of radical immanence, was never following animals. The rethought concept of the animal does not primarily represent a new way of thinking animals, but a new way of thinking altogether. Of course, it means that animals are rethought as well; however, not as a specific case, but along the lines of a more general question – what can they, or their bodies (although the possessive pronoun seems a bit out of place here), do? This is not a question of usefulness – as in: "what can an animal do for us?" – but one of affective capabilities. Posing the question like that naturally leads to including animals in ethical or political communities; however, the character of those communities is, as shown in the last chapter, radically different from what we know.

While rethinking man has been definitely one of the most important parts of this study, it cannot simply be said that it is trying to preach "death of man" nearly half a century after Foucault's works, or the "death of God" more than a century after Nietzsche. If anything, in this work I tried to show that the opposite is the case. The analysis of Kant's philosophy as grounded in a supposed "atheism" proved that it is not enough to deny God's existence to be an atheist, or to think in terms of a plane of immanence. Current animal studies, especially in the case of the identity approach, are in the same spot in reference to the concept of the human and the animal – they seem to cross out the frontier between men and animals, but in fact keep the same logic that differentiates the animal from the human. God is not dead, and neither is man – in fact, one might say that philosophers have faked those deaths, while actually reinforcing the reign of God and man over our thinking.

This is especially visible in the problematic and nature of the content of the previous chapter. Taken without the process that lead to them, its results seem perplexing if not downright absurd – thinking stones, ethics without commandments or the notions of good or evil, communities of hands, etc. This unease is not caused by the peculiarity of philosophical language – in fact, the chapter is not extremely technical – nor by the will to shock or stir controversy with the use of unfounded novelty, as the predecessors of this kind of thinking are clearly traced and the scientific examples used are well founded and known. The reason is far simpler: the concepts – starting from the concept of the animal – after their recreation on a radically different plane, seem to have lost any connection with how they were traditionally used; in fact, they seem to be connected to the old ones in name only.

This is partly true. We are so used to thinking in terms of the plane of transcendence that radical immanence without transcendence seems incomprehensible. Nietzsche described a similar effect in one of the most known fragments of *Twilight of the Idols*:

5.   The "true world" – an idea that is of no further use, not even as an obligation, – now an obsolete, superfluous idea, consequently a refuted idea: let's get rid of it! (Bright day; breakfast; return of *bon sens* and cheerfulness; Plato blushes in shame; pandemonium of all freespirits.)

6.   The true world is gone: which world is left? The illusory one, perhaps? . . . But no! we got rid of the illusory world along with the true one![1]

The ideas of the true world and the illusory, sensual one – the world of Platonic ideas, of God or Truth, and that of our everyday experience – were so dependent upon each other that one could not think properly of the sensual world without taking into account the suprasensual one. The sensual world, as we know it, in fact existed only insofar as it was opposed – and therefore linked – to the "true world." One would have to learn to think of the world that is left in terms other than "sensual" or "illusory."

The case is very much the same with the question of immanence or the animal in this work. Without transcendence, one has to completely re-invent immanence; without the human, one must fundamentally recreate the concept of the animal. The same goes for (animal) ethics, thinking, politics, death, etc. In fact, one might argue that one should invent new terms altogether.

As I said, this is partly true, but only partly. While the concepts indeed deeply change their scope and meaning on the new plane, there is an important reason to keep them, which has a historical and an analytical aspect. Historically, it is important to understand that the developments described in this text are inscribed in the history of rethinking the animal and the human, and are as dependent on the tradition of animal studies as they are on that of atheist philosophy (and Western philosophy in general), in which Spinoza plays a key role, regardless of how critical they might be of those traditions.

The analytical aspect of the reason for keeping the concepts – especially the concept of the animal – is much more important. It is only thanks to the analysis of the "classical" concept of the animal, its link to the human, to immanence and transcendence, that it was possible to sketch out the creation of the new concept of the animal. One never creates (concepts or otherwise) in a vacuum: while this sounds like a platitude, it is often ignored. Unworking[2] the concept of the animal to the core of its logic – its link to the concept of the human and the pair immanence/transcendence – helped show the misguided nature of the attempts to simply ignore the radical difference between the animal and the human, or to construct a plane of immanence by negating transcendence, but keeping the logic that produced it.

To end on a more conciliatory note, if there is one fundamental point in which the aims of (critical) animal studies – or, more generally, the various posthumanist philosophies they form a part of – and this work coincide, it is the notion that man has not yet swallowed and digested the blows to his pride he received from science and philosophy. We still tend to think about a certain idea of (human) subjectivity as special, privileged in the contact with the world. However, some theorists seem to be much more optimistic than I am about the chances of changing this state of affairs, thinking that raising knowledge about the theory of evolution and the basic identity between men and animals is enough; or that, indeed, all that we need is to understand that the human/animal divide makes no sense, as there has never been something like the human in the first place. While this is perhaps true, in this work I have tried to work towards a more modest, if fundamental, change – to recreate a philosophical concept that seemed outright illogical from the beginning.

To finish off, I will ask – as some more practically oriented readers may have already – what can practical uses of such a theory be, and in what way does it change (if at all) the current way to go about the ethical and political challenges the pro-animal movement faces? I have – as I indicated in the Introduction – refrained from asking such direct questions before, but it seems that it might be useful to touch upon this problem for two reasons.

The first one concerns the general question of the so-called continental approaches to animal studies, which have more often than not have been accused of practical – ethical or political – uselessness. Derrida seems to be cited especially often in this context – I have already shown Matthew Calarco's and Paola Cavalieri's reservation in Chapter 1. Danielle Sands reiterates this point in a more general and explicit fashion, adding another element to the critique:

> It is clear [. . .] that deconstructive thought does have a political element. What is less clear, however, is how that political thought translates into political praxis. [. . .] Derrida's tendency to consider the constitution of justice rather than law (which remains an abstraction), the conditions of politics

rather than politics itself, ironically means that when he does address the practicalities of law and politics he is less, rather than more, radical.[3]

In other words – and giving this argument a more general character – one might say that while continental thought (at least in the iterations I have been using in this text) can be radical and revolutionary on a theoretical level, it is also quite conservative on a practically political or ethical level. The reason for this – both in Derrida and in the interpretation of Deleuze I have been trying to pursue here – might be that such accounts, more interested in the singular event or the concrete ethical situation than in general moral laws, tend to value prudence and caution over clear-cut distinctions that seem to be needed in a revolutionary movement. By refusing to annul the singularity of each ethical and political encounter, they might seem more conservative than, for example, radical abolitionist theorists would like them to be.

The second reason for asking about the practical implications of the theory developed here is linked to the particular choice of an interpretation of Spinoza that is quite close to Deleuze's. In the context of the animal problem, it is usually different texts and concepts that are invoked – especially that of *becoming-animal*. Matthew Calarco, in his survey of theoretical approaches in animal studies, sees what he calls the "indistinction" approach – in the case of Deleuze, based primarily on a reading of the book on Bacon – as the most promising practical strategy.[4] And even here, Deleuze is only a part of the argument, empowered by the likes of Agamben and Val Plumwood.

Focusing on Deleuze's reading of Spinoza and ethics as ethology seems to be an even less promising avenue. I have already pointed out that this reading seeks to do away with any sort of morality understood as the formulation of laws or duties, and points rather to experimentation, uncertainty and readiness to affect and especially be affected in as many ways as possible. Any "commandments" one might draw from such an ethics – for example, the one to "open oneself to the undetermined, hidden possible worlds that are expressed in the affective signs of the other,"[5] as advised by the already-cited Ronald Bogue – are always suggested with great reserve. If anything, the amorality of this ethical vision seems to easily degrade into a callous "anything goes" attitude towards both human and animal others.

I do not believe this to be the case. While choosing to practice ethics as ethology – or the politics that go alongside them – does not entail clear-cut laws and duties, it does at the least preclude certain choices. It might be important to mention that while such a formulation is negative (defining immanent ethics as something it does not do), it stems from positive practice of immanence – any rejection is a side effect of this approach.

For example, the ethological approach is incongruous with any Kant-inspired rights ethics. On a theoretical level, I have tried to show this in Chapter 5, where I discussed the basis of the practical distancing of humans and animals in the works of Donaldson and Kymlicka or Francione. I argued that this practice stems from a specific understanding of the human/animal caesura, which is based on a misunderstanding of immanence – it moves the caesura, granting animals the right to certain aspects of transcendence, but backfires by designing them as innocent victims in any human–animal contact, even, at least at face value, beneficial, as in the case of well-cared-for pets, which should (at least according to Francione) cease to exist altogether.

Here I would like to add that this particular element of abolitionist views – and especially its justification – is also at odds with the approach presented in this text. Any type of being open to the affects of others, and acting so as to multiply one's own affective capabilities, needs to be funded on some kind of contact. It is not the place here to give a detailed account of the types of such contact – but figuring this out would require a combination of knowledge and prudence; it might well turn out that in some cases a radical leaving-alone is in order, but the reason should not be guilt or a romantic view of nature. In any way, the ethics-as-ethology approach is certainly closer in this aspect to Haraway, who urges for species to "meet" and pursue their mutual co-creation, than to Francione. It is interesting to note that what Haraway often endeavors in her texts – especially the *Companion Species Manifesto*[6] – can be read as exactly this kind of affective analysis that the approach sketched out in the previous chapter proposes (with some theoretical caveats I have previously mentioned). What bears repeating is that Haraway – to use the language of this analysis – takes into account a wide variety of affects, not only biological, but also socio-political, and treats them on a level playing field.

A more difficult problem is raised by radical abolitionists such as MacCormack. As noted already in Chapter 5, MacCormack's call for self-assured extinction or suicide of the human – a radical distancing of the human and the animal – is based on a critique of the human (or of transcendence) quite akin to what has been endeavored in this study. It is also presented – and defended – in terms of affects:

> Suicide is a recombining of chaos potentials that results in waves of particle affects which precede and exceed the tentative myth of absence/presence but that comes from a certain will to a new occupation of cosmic consistency. Vitalistic suicide is not a cop out, nor is living necessarily a choice to be a certain kind of subject. Life continues after suicide through affect. [. . .] To die is to die actively, to live to affect others for the express purpose of affecting others so they may live vitalistically.[7]

What MacCormack seems to be arguing – here and in the fragments cited in Chapter 5 – human extinction as humans will in fact mean fulfilling the "duty" proposed by Bogue, the duty of opening ourselves to affecting and being affected. After human extinction, after suicide, there will be more rather than less affect (and of the good kind). Judging from the immense damage people have done to the environment, eliminating individuals and species (and their respective affective capabilities), and the reluctance to stop moving this way, such a statement may be tempting. I believe, however, that there are two reasons to be skeptical of such an argument. Firstly, and on a general level, MacCormack seems to be quite pessimistic about the affective capabilities of humans as such and prefers to annihilate them altogether than to work to multiply them. Secondly, and on an individual level, not only suicide is not very Spinozian – as "a free man [. . .] thinks of death least of all things, and his wisdom is a meditation upon life"[8] – but it also includes making death a goal of life, dangerously closing in on Heideggerian territory and the links to transcendence lurking therein. In this way, this approach is at odds with the one I presented throughout this book, and especially in the closing parts of last chapter.

Aside from different forms of Kantian deontology and abolitionist approaches, the ethics-as-ethology approach is also slightly at odds – albeit for a different reason – with the other key ethical theory usually employed to the problem of the animal, namely utilitarianism. The founding moment for the utilitarian approach to animals is of course Jeremy Bentham's famous statement about animal suffering as the basic ethical problem, rather than reason, speech or the like.[9] On the level of the analysis of affects, this is certainly praiseworthy – what Bentham does is to include another affect of a specific type of "other" in our judgment, thus raising our capabilities to affect and be affected in the right way. The problem with utilitarianism – from the point of view of ethics of immanence – lies elsewhere. Firstly, it is based on a very individualistic notion of the (animal or human) subject, which precludes any thinking in terms of variable political communities we have seen in the preceding chapter (and with the example of von Uexküll's oak). Secondly, in its various iterations it is usually based only on a subset of affects – for example, in Singer's version, only "preferences."

I would like to end this analysis of the practical uses of the theory of ethics and politics of immanence with two instances of its positive use. The first one is a question I have consciously put aside for most of this book, namely the relationship between the theoretical and practical treatment of the animal and of women, racial or ethnic minorities, people with mental or physical disabilities or other humans not conforming to the model of what Rosi Braidotti refers to using the image of the Vitruvian Man. This relationship has been well established, and has been the subject of many books[10] and critiques of authors who fail to take it into consideration, such as Agamben.[11] Since my analysis focused on the concept of man – paradigmatically white, male, etc. – I cannot add much to this growing field, but I believe that the affects-based approach advocated here might be of some use for it.

Reflections gathered in collections such as A. Breeze Harper's *Sistah Vegan* or Lisa Kemmerer's *Sister Species* – books which focus not on animal rights philosophies or theories, but on individual stories of women who are Black vegans (in the former case) or animal activists (in the latter) – show that real-life decisions are rarely made according to any utilitarian- or rights-based approach; they are a messy conglomerate of individual and family influences, the understanding of history, social factors such as gender and race, actual contacts with animals and so on, showing the importance of actual lived experience in those choices – something professional philosophers often seem to miss. Indeed, as Martha Nussbaum already noticed, even the most influential books on the animal problem – such as Singer's *Animal Liberation* – might owe a lot of their success not so much on the soundness of the theories, but on the imaginative qualities of the writing: "Good imaginative writing has been crucial in motivating opposition to cruelty toward animals."[12]

But – even to a philosopher – those books offer something more than simply individual stories. They show that there is indeed no telling which affects will turn out decisive in motivating which attitudes toward animals, getting across the point that we do not, indeed, know, what a body can or will do with what it has been affected with. Let me give two examples, both from A. Breeze Harper's collection.

Michelle R. Lloyd-Page writes:

> Seeing a connection between the treatment of feed animals, laying chick-ens, and people of color is a rather recent phenomenon for me. Two years ago, I wouldn't have believed there was such a connection. Today, I know better. The connection becomes clear with a careful reading of our history and an understanding of the true nature of food production in the United States. The connection, however, is also observable by a thorough analysis of today's headlines and an informed critique of social policy and commu-nity life.[13]

To the plethora of ways of seeing the connection between the treatment of farm animals and people of color, one must add that the intensity of seeing the con-nection was only possible for Lloyd-Page (herself a professor of sociology) – according to her own account – thanks to what she calls a *kairos* moment in a KFC restaurant. None of these things are more or less important to her account of transitioning to a vegan diet (or at least not essentially so). What this shows is a multitude of affects – social, historical, political, visual, communal, scientific and mundane – that are at work in every situation. Any general theory – especially a philosophical ethics of one or another sort – would deem at least some of these irrelevant: history has little to offer to a utilitarian who urges us to calmly count satisfied and frustrated preferences of the parties involved; *kairos* moments have little to offer someone who focuses on social causality. For ethics-as-ethology, combining all these affects on one plane is perfectly permissible – and so is searching for others which might further influence the situation.

The second example shows a corollary to the first, analyzing not the various causes, but various outcomes (much as this distinction is in itself a simplification). Layli Phillips, even though she is a radical vegan herself, has a positive attitude to – as she calls it – "sloppy" veganism:

> [M]any vegans refrain from eating meat, dairy, and eggs, yet eat honey or wear leather. Other vegans shop vegan and eat vegan at home but look the other way at a vegetarian restaurant for dishes that use a small amount of but-ter, cream, or cheese. Some vegans may take a bite of cake that contains eggs at the party of a really good friend who isn't vegan. Some vegans are vegan everywhere except at their grandmother's house![14]

Again, the multitude of factors that influence the choices of even professed vegans is impossible to be adequately subsumed under classical "rights" or utilitarian approaches, and shows the multitude of affects and outcomes that are in play when food choices are concerned.

One remark seems in order – as said before, while ethics inspired by Spinoza does not include the notions of Good and Evil, it allows the discussion of better and worse outcomes, and this gives it critical power. This power should certainly be exercised in any affective tally. To give just one example, the choice to adopt a vegan diet – as evidenced by some of the essays in A. Breeze Harper's collection – is often influenced by the general idea of purity, which often includes the refusal of vaccination or of standard medical treatments. It is not a place here to discuss the hypothetical question "What would Spinoza say?" (in any way, such a staunch rationalist with high esteem for science hardly seems like the type to subscribe to what the UN euphemistically calls "vaccine hesitancy"), but any notion of "purity" (or "naturalness") is certainly problematic from the point of view of immanence, for it seems to recognize a transcendent ideal or referent that is incongruous with the pragmatic, material nature of ethics-as-ethology.

The last use of the approach proposed in this book I would like to suggest has more to do with our attitude to animals themselves. As Carol J. Adams points out, one of the processes thanks to which we can stomach the abhorrent treatment of animals, especially those industrially farmed – as well as human-to-human atrocities such as genocide – is the linguistic practice of turning them into mass terms. As Adams notes,

> Mass terms are linked to subjects being diminished [. . .] [a]fter genocides or fratricides like the Civil War, the survivors dig up the bodies buried in mass graves, as in Rwanda or Gettysburg, and try to reassert through separate burials each one's individuality against the annihilation of the mass term. We cannot undo the act of genocide – the dead are dead – but we can undo part of the idea that allows genocide, the use of mass terms, by asserting the individual and maintaining our ties to the dead as individuals. [. . .] Animals are killed because they are false mass terms, but they die as individuals – as a cow, not as beef; as a pig, not as pork. Each suffers his or her own death, and this death matters a great deal to the one who is dying.[15]

What is to be done, according to Adams, is for the animals to become once again individualized, in a way that people are individualized – a similar recipe to the one Leonard Lawlor proposes in his reading of Derrida, to which I alluded (and with which I disagreed) in Chapter 1.

I have already implicitly problematized such individualization in Chapter 5. At least on a theoretical level, it would amount to humanizing the animal, which would amount to simply moving the human–animal caesura from one place to another, without actually problematizing the two notions and creating a plane of transcendence. On a strictly practical plane, I would argue that while such an individualization seems to have been proven to be helpful in the case of humans – and it is certain that our "individualized" home animals are treated much better than those who have been unfortunate enough to become a referent of the "mass term" (Deleuze's reservations notwithstanding) – it might not be sufficient. The scope of this chapter is not enough to give a full critique of the problem, but let me just point to the fact that this individualization would have to proceed by some kind of imaginative compassion (akin, perhaps, to the "sympathetic imagination" proposed by Coetzee's Elizabeth Costello),[16] which many note to be a severely limited faculty when it comes to big numbers of beings[17] – and with factory farmed animals, we are dealing with staggering numbers indeed.

What the immanence approach proposes is another exercise of imagination altogether. As suggested in the "Political Animal" part of Chapter 6, rather than imagining every animal (or human) as singular, this approach allows us to change focuses between groupings, multiplicities and communities. To remain only on the plane of individual vs. mass is to lose possibilities offered by the immanence-based idea that not only are individuals themselves composed of different parts forming different communities, but that those individuals and their parts form any number of communities that broach each-other's boundaries in unexpected ways. This is not only theoretically interesting and ethically illuminating, but also helps to see and understand various possible political alliances, in turn allowing all parties to experiment with new ways to affect and be affected.

This, of course, is a maximalist and radical program, and any endeavors to put it in place should be exercised with the same caution that pertains to any ethico-ethological experiment; in particular, it should not form an easy excuse to light-heartedly abandon any of the already-existing pro-animal program. The ethics and politics of immanence are perhaps best understood according to a Deleuzian strategy of combination and conjunction (and . . . and . . . and. . .) rather than replacement. In any case, it seems that – to paraphrase Deleuze and Guattari – we should never believe that the plane of immanence alone will suffice to save us.

Notes

1 Friedrich Nietzsche, "Twilight of the Idols, or How to Philosophize with a Hammer," in *The Anti-Christ, Ecce Homo, Twilight of the Idols*, trans. Judith Norman (Cambridge: Cambridge University Press, 2005), 171.
2 For obvious reasons I refrain from using the term "deconstruction" in this context.
3 Danielle Sands, " 'Beyond' the Singular? Ecology, Subjectivity, Politics," in *The Animal Catalyst: Towards Ahuman Theory*, ed. Patricia MacCormack (London: Bloomsbury, 2014), 63.
4 Matthew Calarco, *Thinking Through Animals: Identity, Difference, Indistinction* (Stanford: Stanford Briefs, 2015), 57–59.
5 Ronald Bogue, "Immanent Ethics," in *Deleuze's Way: Essays in Transverse Ethics and Aesthetics* (Aldershot: Ashgate, 2007), 13.
6 Donna Haraway, "The Companion Species Manifesto: Dogs, People and Significant Otherness," in *Manifestly Haraway* (Minneapolis and London: University of Minnesota Press, 2016).
7 Patricia MacCormack, "After Life," in *The Animal Catalyst*, 178.
8 Baruch Spinoza, "Ethics," in *Complete Works*, trans. Samuel Shirley (Indianapolis: Hackett, 2002), 355.

9 Jeremy Bentham, "An Introduction to the Principles of Morals and Legislation," in *The Works of Jeremy Bentham* (Bristol: Thoemmes Press, 1995), vol. 1, 143.

10 Examples include Carol J. Adams, *The Sexual Politics of Meat: A Feminist-Vegetarian Critical Theory* (London: Bloomsbury Academic, 2015); Lisa A. Kemmerer, ed., *Sister Species: Women, Animals and Social Justice* (Chicago: University of Illinois Press, 2011) or A. Breeze Harper, ed., *Sistah Vegan: Black Female Vegans Speak on Food, Identity, Health, and Society*, Kindle ed. (New York: Lantern Books, 2009).

11 See, e.g., Ewa Plonowska Ziarek, "Bare Life on Strike: Notes on the Biopolitics of Race and Gender," *South Atlantic Quarterly* 107, no. 1 (Winter 2008): 89–105.

12 Martha Nussbaum, *Frontiers of Justice: Disability, Nationality, Species Membership* (Cambridge: The Belknap Press, 2007), 354.

13 Michelle R. Lloyd-Page, "Thinking and Eating at the Same Time: Reflections of a Sistah Vegan," in *Sistah Vegan*.

14 Layli Philipps, "Veganism and Ecowomanism," in *Sistah Vegan*.

15 Carol J. Adams, "The War on Compassion," in *The Animal Catalyst*, 18–19.

16 See J. M. Coetzee, *The Lives of Animals* (Princeton: Princeton University Press, 1999).

17 A useful (if brief) discussion of this problem with regard to people (and the question of cosmopolitanism) is provided in Elaine Scarry, "The Difficulty of Imagining Other People," in *For Love of Country?* ed. Martha Nussbaum (Boston: Beacon Press, 2002), 98–110.

Bibliography

Adams, Carol J. *The Sexual Politics of Meat: A Feminist-Vegetarian Critical Theory*. London: Bloomsbury Academic, 2015.

Agamben, Giorgio. *Infancy and History*. Translated by Liz Heron. London: Verso, 1993.

Agamben, Giorgio. *The Open: Man and Animal*. Translated by Kevin Attell. Stanford: Stanford University Press, 2004.

Aquinas, Thomas. *The "Summa Theologica"*. Translated by The Fathers of the English Dominican Province. London: Burns Oates & Washbourne, 1942.

Baltas, Aristides. *Peeling Potatoes or Grinding Lenses: Spinoza and Young Wittgenstein Converse on Immanence and Its Logic*. Pittsburgh: University of Pittsburgh Press, 2012.

Barad, Judith. *Aquinas on the Nature and Treatment of Animals*. London and San Francisco: International Scholars Publications, 1995.

Bataille, Georges. *Theory of Religion*. Translated by Robert Hurley. New York: Zone, 1989.

Beistegui, Miguel de. *Immanence: Deleuze and Philosophy*. Edinburgh: Edinburgh University Press, 2010.

Bentham, Jeremy. "An Introduction to the Principles of Morals and Legislation." In *The Works of Jeremy Bentham*, vol. 1. Bristol: Thoemmes Press, 1995, 1–154.

Bernasconi, Robert. "No Exit – Levinas' Aporetic Account of Transcendence." *Research in Phenomenology* 35 (2005): 101–17.

Bobiec, Andrzej, ed. *The Afterlife of a Tree*. Warszawa and Hajnówka: WWF Poland, 2005.

Bogue, Ronald. "Immanent Ethics." In *Deleuze's Way: Essays in Transverse Ethics and Aesthetics*. Aldershot: Ashgate, 2007, 7–16.

Braidotti, Rosi. *The Posthuman*. Cambridge: Polity Press, 2013.

Braidotti, Rosi. *Transpositions: On Nomadic Ethics*. Cambridge: Polity Press, 2006.

Buben, Adam. *Meaning and Mortality in Kierkegaard and Heidegger*. Evanston: Northwestern University Press, 2016.

Buchanan, Brett. *Onto-Ethologies: The Animal Environments of Uexküll, Heidegger, Merleau-Ponty and Deleuze*. Albany: SUNY Press, 2008.

Butler, Rex. *Deleuze and Guattari's 'What Is Philosophy'*. London and New York: Bloomsbury, 2016.

Calarco, Matthew. "Identity: Difference. Indistinction." *CR: The New Centennial Review* 11, no. 2 (2011): 41–60.

Calarco, Matthew. *Thinking Through Animals: Identity, Difference, Indistinction*. Stanford: Stanford Briefs, 2015.

Calarco, Matthew. *Zoographies. The Question of the Animal from Heidegger to Derrida*. New York: Columbia University Press, 2008.

Cassirer, Ernst. *An Essay on Man: An Introduction to a Philosophy of Human Culture*. New York: Doubleday, 1956.

Castricano, Jodey, ed. *Animal Subjects*. Waterloo, Ontario: Wilfrid Laurier University Press, 2008.

Cavalieri, Paola. *The Animal Question: Why Nonhuman Animals Deserve Human Rights*. Translated by Catherine Woollard. New York: Oxford University Press, 2001.

Ceronetti, Guido. *The Silence of the Body: Materials for the Study of Medicine*. Translated by Michael Moore. New York: Farrar, Strauss and Giroux, 1993.

Clark, David L. "Kant's Aliens: The Anthropology and its Others." *CR: The New Centennial Review* 1, no. 2 (Fall 2011): 201–89.

Colebrook, Claire. *Deleuze and the Meaning of Life*. London and New York: Continuum, 2010.

Dallmayr, Fred. "An 'Inoperative' Global Community? Reflections on Nancy." In *On Jean-Luc Nancy. The Sense of Philosophy*, edited by Darren Sheppard, Simon Sparks and Colin Thomas. London and New York: Routledge, 1997, 168–89.

Dalton, Stuart. "Nancy and Kant on Inoperative Communities." *Critical Horizons* 1, no. 1 (February 2000): 29–50.

Davies, Paul. "Sincerity and the End of Theodicy: Three Remarks on Levinas and Kant." In *The Cambridge Companion to Levinas*, edited by Simon Critchley and Robert Bernasconi. Cambridge: Cambridge University Press, 2004.

Dawkins, Richard. *The Selfish Gene*. Oxford: Oxford University Press, 2006.

Deleuze, Gilles. *Difference and Repetition*. Translated by Paul Patton. New York: Columbia University Press, 1994.

Deleuze, Gilles. *Expressionism in Philosophy: Spinoza*. Translated by Martin Joughin. New York: Zone, 1990.

Deleuze, Gilles. *Francis Bacon: The Logic of Sensation*. Translated by Daniel W. Smith. London and New York: Continuum, 2004.

Deleuze, Gilles. *Nietzsche and Philosophy*. Translated by Hugh Tomlinson. London and New York: Continuum, 1993.

Deleuze, Gilles. *Proust and Signs*. Translated by Richard Howard. Minneapolis: University of Minnesota Press, 2000.

Deleuze, Gilles. *Spinoza: Practical Philosophy*. Translated by Robert Hurley. San Francisco: City Light Books, 1988.

Deleuze, Gilles. "To Have Done with Judgment." In *Essays Critical and Clinical*, translated by Daniel W. Smith and Michael A. Greco. London: Verso, 1998, 126–35.

Deleuze, Gilles. *Two Regimes of Madness*. Translated by Ames Hodges and Mike Taormina. New York: Semiotext(e), 2006.

Deleuze, Gilles and Felix Guattari. *A Thousand Plateaus*. Translated by Brian Massumi. Minneapolis: University of Minnesota Press, 2005.

Deleuze, Gilles and Felix Guattari. *What Is Philosophy?* Translated by Hugh Tomlinson and Graham Burchell. New York: Columbia University Press, 1994.

Deleuze, Gilles and Claire Parnet. *Dialogues*. Translated by Hugh Tomlinson and Barbara Habberjam. New York: Columbia University Press, 1987.

Derrida, Jacques. *The Animal that Therefore I Am*. Translated by David Wills. New York: Fordham University Press, 2008.

Derrida, Jacques. *Aporias: Dying – Awaiting (one another at) the Limits of Truth*. Translated by Tomas Dutoit. Stanford: Stanford University Press, 1993.

Derrida, Jacques. *The Beast and the Sovereign. Volume I*. Translated by Geoffrey Bennington. Chicago: University of Chicago Press, 2009.

Derrida, Jacques. "*Geschlecht* II. Heidegger's Hand." In *Deconstruction and Philosophy: The Texts of Jacques Derrida*, edited by John Sallis. Chicago and London: University of Chicago Press, 1987, 161–96.

Derrida, Jacques. *The Politics of Friendship*. Translated by George Collins. London: Verso, 2005.

Derrida, Jacques and Jean-Luc Nancy. "Eating Well." In *What Comes After the Subject?* edited by Eduardo Cadava, Peter Connor, and Jean-Luc Nancy. London and New York: Routledge, 1991, 96–119.

Derrida, Jacques and Elizabeth Roudinesco. "Violence Against Animals." In *For What Tomorrow: A Dialogue*. Translated by Jeff Fort. Stanford: Stanford University Press, 2004, 62–76.

Descartes, René. "Meditations on First Philosophy." In *The Philosophical Works of Descartes*, translated by Elizabeth S. Haldane and G. R. T. Ross. Cambridge: Cambridge University Press, 1967.

Descartes, René. *The Philosophical Writings of Descartes, Vol III: The Correspondence*. Translated by John Cottingham, et al. Cambridge: Cambridge University Press, 1991.

Descartes, René. *Principles of Philosophy*. Translated by John Veitch. Whitefish: Kessinger, 2004.

Donaldson, Sue and Will Kymlicka. *Zoopolis: A Political Theory of Animal Rights*. Oxford: Oxford University Press, 2011.

Dosse, Francois. *Gilles Deleuze and Felix Guattari: Intersecting Lives*. Translated by Deborah Glassman. New York: Columbia University Press, 2010.

Dreyfuss, Hubert and Mark Rathall, eds. *Heidegger Reexamined, Vol. 1. Dasein, Authenticity and Death*. New York: Routledge, 2002.

Dreyfuss, Hubert and Mark Rathall, eds. *Heidegger Reexamined, Vol. 4. Language and the Critique of Subjectivity*. New York: Routledge, 2002.

Fontenay, Elizabeth de. *Le silence de bêtes*. Paris: Fayard, 1998.

Foster, Nicola L. and Mark Briffa. "Familial Strife on the Seashore: Aggression Increases with Relatedness in the Sea Anemone *Actinia equina*." *Behavioural Processes* 103 (March 2014): 243–45.

Francione, Gary L. and Anna E. Charlton, "The Case Against Pets." *Aeon*, September 8, 2016. https://aeon.co/essays/why-keeping-a-pet-is-fundamentally-unethical.

Freud, Sigmund. "Group Psychology and the Analysis of the Ego." In *Standard Edition of the Complete Psychological Works of Sigmund Freud, Volume XVIII: Beyond the Pleasure Principle, Group Psychology and Other Works*, edited by James Strachey. London: The Hogarth Press, 1955, 65–144.

Freud, Sigmund. "Introductory Lectures on Psychoanalysis (part III)." In *Standard Edition of the Complete Psychological Works of Sigmund Freud, Vol. XVI*, edited by James Strachey. London: The Hogarth Press, 1981, 243–463.

Freud, Sigmund. *Totem and Taboo*. Translated by James Strachey. London and New York: Routledge, 2004.

Gay, Peter, ed. *The Freud Reader*. New York: W. W. Norton and Company, 1989.

Gontier, Thierry. *De l'homme à l'animal. Montaigne et Descartes ou paradoxes de la philosophie moderne sur la nature des animaux*. Paris: Vrin, 1998.

Hagglund, Martin. *Radical Atheism: Derrida and the Time of Life*. Stanford: Stanford University Press, 2008.

Haraway, Donna. "The Companion Species Manifesto: Dogs, People and Significant Otherness." In *Manifestly Haraway*. Minneapolis and London: University of Minnesota Press, 2016, 91–198.

Haraway, Donna. *When Species Meet*. Minneapolis: University of Minnesota Press, 2008.

Harper, A. Breeze, ed. *Sistah Vegan: Black Female Vegans Speak on Food, Identity, Health, and Society*. Kindle ed. New York: Lantern Books, 2009.

Hatab, Lawrence J. "From Animal to Dasein: Heidegger and Evolutionary Biology." In *Heidegger on Science*, edited by Trish Glazebrook. Albany: State University of New York Press, 2012, 93–111.

Heidegger, Martin. *Being and Time*. Translated by John Macquarrie and Edward Robinson. New York: Harper and Row, 1962.

Heidegger, Martin. *The Fundamental Concepts of Metaphysics: World, Finitude, Solitude*. Translated by William McNeill and Nicholas Walker. Bloomington and Indianapolis: Indiana University Press, 1995.

Heidegger, Martin. *Pathmarks*. Edited by William McNeill. Cambridge: Cambridge University Press, 1998.

Heidegger, Martin. "The Thing." In *Poetry, Language, Thought*, translated by Albert Hofstadter New York: Harper Perennial, 2001), 161–84.

Heidegger, Martin. *What Is Called Thinking*. Translated by Fred D. Wieck and J. Glenn Gray. London, New York and Evanston: Harper and Row, 1968.

Heinrich, Bernd. *Life Everlasting: The Animal Way of Death*. Boston: Houghton Mifflin Harcourt, 2012.

Injaian, Allison and Elizabeth A. Tibbetts. "Cognition Across Castes: Individual Recognition in Worker Polistes Fuscatus Wasps." *Animal Behaviour* 87 (January 2014): 91–96.

Kant, Immanuel. "Anthropology from a Pragmatic Point of View." In *Anthropology, History, and Education*, translated by Robert B. Louden. Cambridge: Cambridge University Press, 2007, 227–429.

Kant, Immanuel. *Critique of the Power of Judgment*. Translated by Paul Guyer and Eric Matthews. Cambridge: Cambridge University Press, 2000.

Kant, Immanuel. *Critique of Pure Reason*. Translated and edited by P. Guyer and Allen W. Wood. Cambridge: Cambridge University Press, 1998.

Kant, Immanuel. *Lectures on Ethics*. Translated by Louis Infield. New York: Harper and Row, 1963.

Kant, Immanuel. *Practical Philosophy*. Translated and edited by Mary J. Gregor. Cambridge: Cambridge University Press, 1999.

Kemmerer, Lisa A., ed. *Sister Species: Women, Animals and Social Justice*. Chicago: University of Illinois Press, 2011.

Korsgaard, Christine. "Fellow Creatures: Kantian Ethics and Our Duties to Animals." Tanner Lecture on Human Values delivered on February 6, 2004. Accessed February 3, 2018. https://tannerlectures.utah.edu/_documents/a-to-z/k/korsgaard_2005.pdf.

Lacan, Jacques. *Écrits: A Selection*. Translated by Alan Sheridan. New York: Norton, 1977.

Laertius, Diogenes. *Lives of Eminent Philosophers*. Translated by Robert Drew Hicks. Cambridge, MA and London: Harvard University Press/William Heineman, 1956.

Lane, Melissa S. *Method and Politics in Plato's Statesman*. Cambridge: Cambridge University Press, 1998.

Latour, Bruno. *We Have Never Been Modern*. Cambridge, MA: Harvard University Press, 1993.

Lawlor, Leonard. *This Is Not Sufficient: An Essay on Animality and Human Nature in Derrida*. New York: Columbia University Press, 2007.

Levinas Emmanuel. "The Other Transcendence." In *Alterity and Transcendence*, translated by Michael B. Smith. London: The Athlone Press, 1999, 1–76.

Levinas, Emmanuel. *Otherwise than Being*. Translated by Alphonso Lingis. Dodrecht: Kluwer, 1991.

Levinas, Emmanuel. *Totality and Infinity*. Translated by Alphonso Lingis. The Hague: Martinus Nijhoff, 1979.

Lingis, Alphonso. "Animal Body, Inhuman Face." In *Zoontologies: The Question of the Animal*, edited by Cary Wolfe. Minneapolis and London: University of Minnesota Press, 2003, 165–82.

Linnaeus, Carolus [Carl von Linné]. *Menniskans Cousiner*. Edited by Telemak Fredbärj. Uppsala: Ekenäs, 1955.

Linzey, Andrew. *Animal Theology*. Chicago: University of Illinois Press, 1994.

Linzey, Andrew. "Christianity and the Rights of Animals." *The Animals' Voice* 2, no. 4 (August 1989).

MacCormack, Patricia, ed. *Towards Ahuman Theory*. London: Bloomsbury, 2014.

Margulis, Lynn. *Symbiotic Planet*. New York: Basic Books, 1999.

Massumi, Brian. *What Animals Teach Us About Politics*. Durham and London: Duke University Press, 2014.

Montag, Warren. "Interjecting Empty Spaces: Imagination and Interpretation in Spinoza's Tractatus Theologico-Politicus." In *Spinoza Now*, edited by Dimitris Vardoulakis. Minneapolis: University of Minnesota Press, 2011, 161–78.

Montag, Warren and Ted Stolze, eds. *The New Spinoza*. Minneapolis: University of Minnesota Press, 1997.

Montaigne, Michel de. *The Complete Essays*. Translated by Michael Andrew Screech. London: Penguin, 2003.

Nancy, Jean-Luc. "Of Being In-common." In *Community at Loose Ends*, edited by Miami Theory Collective. Minneapolis: University of Minnesota Press, 1991, 1–12.

Nancy, Jean-Luc. *The Inoperative Community*. Translated by Peter Connor, et al. Minneapolis: University of Minnesota Press, 1991.

Negri, Antonio. *The Savage Anomaly*. Translated by Michael Hardt. Minneapolis: University of Minnesota Press, 1991.

Negri, Antonio. *Spinoza for our Time: Politics and Postmodernity*. Translated by William McCuaig. New York: Columbia University Press, 2013.

Negri, Antonio. *Subversive Spinoza. (un)Contemporary Variations*. Edited by Timothy Murphy. Manchester: University of Manchester Press, 2004.

Nietzsche, Friedrich. *Thus Spoke Zarathustra*. Translated by Adrian Del Caro. Cambridge: Cambridge University Press, 2006.

Nietzsche, Friedrich. "Twilight of the Idols, or How to Philosophize with a Hammer." In *The Anti-Christ, Ecce Homo, Twilight of the Idols*, translated by Judith Norman. Cambridge: Cambridge University Press, 2005, 153–230.

Noë, Alva. *Out of Our Heads*. Kindle ed. New York: Hill and Wang, 2010.

Nussbaum, Martha. *Frontiers of Justice. Disability, Nationality, Species Membership*. Cambridge: The Belknap Press, 2007.

Oberst, Joachim L. *Heidegger on Language and Death: The Intrinsic Connection in Human Existence*. London and New York: Continuum, 2009.

Oliver, Kelly. *Animal Lessons: How They Teach Us to Be Human*. New York: Columbia University Press, 2009.

Pignarre, Philippe and Isabelle Stengers. *Capitalist Sorcery: Breaking the Spell*. London: Palgrave Macmillan, 2011.

Plato. "The Statesman." In *The Being of the Beautiful*. Translated and edited by Seth Bernadete. Chicago and London: University of Chicago Press, 1984.

Plonowska Ziarek, Ewa. "Bare Life on Strike: Notes on the Biopolitics of Race and Gender." *South Atlantic Quarterly* 107, no. 1 (Winter 2008): 89–105.

Poirier, Jean-Louis. "Éléments pour un zoologie philosophique." *Critique*, no. special, "L'Animalité" aout-septembre (1978): 673–88.

Powell, Jeffrey, ed. *Heidegger and Language*. Indianapolis: Indiana University Press, 2013.

Rachels, James. *Created from Animals: The Moral Implications of Darwinism*. Oxford and New York: Oxford University Press, 1990.

Regan, Tom. *Animal Rights, Human Wrongs*. Lanham: Rowman & Littlefield, 2003.

Regan, Tom. "The Case for Animal Rights." In *In Defense of Animals*, edited by Peter Singer. New York: Basil Blackwell, 1985, 13–26.

Rey, Jean-Francois. *La mesure de l'homme. L'idée de l'humanité dans la philosophie d'Emmanuel Levinas*. Paris: Michalon, 2001.

Rosanvallon, Jérôme. *Deleuze & Guattari à vitesse infinie: Volume 1, De la vitesse infinie de l'être . . .* Paris: Ollendorff & Desseins, 2009.

Ryle, Gilbert. *The Concept of Mind*. London: Routledge, 2009.

Sauvagnargues, Anne. *Deleuze and Art*. Translate by Samantha Bankston. London: Bloomsbury, 2005.

Scarry, Elaine. "The Difficulty of Imagining Other People." In *For Love of Country?* edited by Martha Nussbaum. Boston: Beacon Press, 2002, 98–110.

Schmitt, Carl. *The Concept of the Political*. Expanded ed. Translated by George Schwab. Chicago: University of Chicago Press, 2007.

Schmitt, Carl. *Political Theology: Four Chapters on the Concept of Sovereignty*. Translated by George Schwab. Chicago: University of Chicago Press, 2005.

Sharp, Hasana. *Spinoza and the Politics of Renaturalization*. Chicago: University of Chicago Press, 2011.

Simondon, Gilbert. *Two Lessons on Animal and Man*. Translated by Drew S. Burk. Minneapolis: Univocal, 2011.

Singer, Peter. *Animal Liberation*. New York: Ecco, 2002.

Singer, Peter. "Animals and the Value of Life." In *Matter of Life and Death*, edited by Tom Regan. New York: Random House, 1980.

Spinoza, Baruch. *Complete Works*. Translated by Samuel Shirley. Indianapolis: Hackett, 2002.

Streiter, Anja. "The Community According to Jean-Luc Nancy and Claire Denis." *Film-Philosophy* 12, no. 1 (2008): 49–62.

Tibbetts, Elizabeth A. and Adrian G. Dyer. "Good with Faces." *Scientific American* 309, no. 6 (December 2013): 62–67.

Tonner, Philip. "Are Animals Poor in the World? A Critique of Heidegger's Anthropocentrism." In *Anthropocentrism: Humans, Animals, Environments*, edited by Rob Boddice. Leiden: Brill, 2011, 203–21.

Uexküll, Jakob von. *A Foray into the Worlds of Animals and Men with A Theory of Meaning*. Translated by Joseph D. O'Neil. Minneapolis: University of Minnesota Press, 2010.

Waal, Frans de. *Are We Smart Enough to Know How Smart Animals Are?* New York: W. W. Norton & Company, 2016.

Waal, Frans de. *Primates and Philosophers: How Morality Evolved*. Princeton: Princeton University Press, 2016.

Weil, Kari. *Thinking Animals*. New York: Columbia University Press, 2012.

Wilson, Margaret Dauler. *Ideas and Mechanism: Essays on Early-Modern Philosophy*. Princeton: Princeton University Press, 1999.

Wittgenstein, Ludwig. *Tractatus Logico-Philosophicus*. Translated by D. F. Pears & B. F. McGuinness. London and New York: Routledge & K. Paul, 1961.

Wohlleben, Peter. *The Hidden Life of Trees: What They Feel, How They Communicate – Discoveries from a Secret World*. Translated by Jane Billinghurst. Vancouver: Greystone Books, 2016.

Wolfe, Cary. *Before the Law. Humans and Other Animals in a Biopolitical Frame*. Chicago and London: University of Chicago Press, 2013.

Wright, Tamra, Peter Hughes and Alison Ainley. "The Paradox of Morality: An Interview with Emmanuel Levinas." In *The Provocation of Levinas: Rethinking the Other*, edited by Robert Bernasconi and David Wood. London and New York: Routledge, 1988, 168–80.

Wróbel, Szymon. "Domesticating Animals: Description of a Certain Disturbance." In *The Animals in Us – We in Animals*, edited by Szymon Wróbel. Frankfurt am Main: Peter Lang, 2014, 219–38.

Yovel, Yirmiyahu. *Spinoza and Other Heretics: The Adventures of Immanence*. Princeton: Princeton University Press, 1989.

Ziarek, Krzysztof. "After Humanism: Heidegger and Agamben." *South Atlantic Quarterly* 107, no. 1 (Winter 2008): 187–209.

Ziarek, Krzysztof. *Language After Heidegger*. Indianapolis: Indiana University Press, 2013.

Zourabichvili, Francois. *Deleuze: A Philosophy of the Event Together with The Vocabulary of Deleuze*. Translated by Kieran Aarons. Edinburgh: Edinburgh University Press, 2012.

Index

Page numbers in *italic* indicate a figure and page numbers in **bold** indicate a table on the corresponding page.

Printed in the United States
by Baker & Taylor Publisher Services

Printed in the United States
by Baker & Taylor Publisher Services